An End to the Upside Down Cosmos

Also by Mark Gober

AN END TO UPSIDE DOWN THINKING

Dispelling the Myth That the Brain Produces Consciousness, and the Implications for Everyday Life

AN END TO UPSIDE DOWN LIVING

Reorienting Our Consciousness to Live Better and Save the Human Species

AN END TO UPSIDE DOWN LIBERTY

Turning Traditional Political Thinking on Its Head to Break Free from Enslavement

AN END TO UPSIDE DOWN CONTACT

UFOs, Aliens, and Spirits—and Why Their Ongoing Interaction with Human Civilization Matters

AN END TO THE UPSIDE DOWN RESET

The Leftist Vision for Society Under the "Great Reset"—and How It Can Fool Caring People into Supporting Harmful Causes

AN END TO UPSIDE DOWN MEDICINE

Contagion, Viruses, and Vaccines—and Why Consciousness Is Needed for a New Paradigm of Health

Note from the author: Throughout, I have slightly edited some quoted material for spelling, punctuation, and clarity. "Emphasis added" is noted when I have added bold or italics in a particular paragraph. If this is not noted, it means that any bold or italics were in the original.

An End to the Upside Down Cosmos

Rethinking the Big Bang, Heliocentrism, the Lights in the Sky…and Where We Live

Mark Gober

Waterside Productions
Cardiff-by-the-Sea, California

The longer-than-normal extract in chapter 4 by Steven A. Young, PhD, from his book *A Fool's Wisdom* (2024), has been printed with his permission.

First Printing, 2024
Printed in the United States of America

ISBN-13: 978-1-962984-55-3: print edition
ISBN-13: 978-1-962984-56-0: ebook edition
ISBN-13: 979-8-228000-61-2: audiobook edition

Waterside Productions
2055 Oxford Avenue
Cardiff-by-the-Sea, CA 92007
www.waterside.com

To those who are comfortable with three important words:
"I don't know."

"The way that science seems to be working these days is, basically, somebody makes something up—it's a construct of their mind—which, since none of us are omniscient or perfect, will eventually be proven to be wrong. Such is the case with all the things that are referred to as 'science': eventually, somebody comes up with an experiment or an observation or a way to demonstrate that the theory about how things work is not true.

"And then the people have a fork in the road. What you would think is that if the new information falsified the claim, then the original theory would be discarded. But that doesn't seem to be how it works in pretty much all of science for the entire twentieth century and into the twenty-first century."

—Thomas Cowan, MD (May 2024)[1]

CONTENTS

PART II

WHAT IS THE NATURE OF OUR HOME?

INTRODUCTION

ERRORS IN PHYSICS AND THINKING

According to modern physics, roughly 96 percent of the cosmos is composed of "dark matter" and "dark energy."[1]

96 percent.

The discovery of dark matter is often attributed to astronomer Fritz Zwicky in 1933.[2] When studying the Coma Cluster of galaxies, he found that it was moving much more quickly than would have been expected based on the laws of gravity (via Albert Einstein's general relativity theory). According to these fundamental principles of physics, there should have been *400 times* as much matter present in order to explain Zwicky's observations.[3] But that extra matter wasn't there. Put another way, the observable matter accounted for *less than 1 percent* of the total mass required.[4] Something was very wrong.

So, scientists have assumed that "dark matter" must have been there to fill in the gap. Otherwise, the observations of the Coma Cluster would not have made sense.[5] The foundations of physics would have needed serious rethinking without dark matter. Therefore, Zwicky's discovery was monumental.

And yet, in spite of dark matter's apparent importance, neither Zwicky nor any modern physicists know what it is. A 2023 *Scientific American* article refers to it as "the invisible substance that serves as gravitational scaffolding for galaxies,"[6] and it is now believed to make up about 27 percent of the cosmos.[7] Astrophysicist Neil deGrasse Tyson calls dark matter "the longest-standing unsolved mystery in astrophysics."[8]

However, Pavel Kroupa, a German astrophysicist and professor at the University of Bonn, noted in a 2023 interview that he thinks dark matter "has been falsified with more than five-sigma confidence [99.99997%]. So, it's not there."[9] [emphasis added]

Kroupa said that he and his collaborators in many countries have been looking at this issue for twenty-five years, and the observations of galaxies predicted by the theory of dark matter are simply not found. He gave an example in the interview:

> If you have galaxies orbiting other galaxies, and they are moving around in the "dark matter halo,"...it will be getting slower and falling towards the center of the other galaxy and merge....So, it's the same [thing] as if you have a pot of honey and you put a marble into the honey. It will not accelerate downwards, but it will slowly sink. But if you have the same pot, without honey, the ball would just fall down quickly....
>
> So basically the same thing [should happen] in a dark matter halo. A galaxy enters this dark matter halo and slowly sinks down. It doesn't accelerate [as] fast as it should if there were no dark matter halo.
>
> **[But] we've tested this observationally and the effect of the slowing down is not there....So galaxies are encountering each other far too rapidly.[10]** [emphasis added]

The interviewer then asked: "So, this was a prediction that was made by the claim that dark matter exists, and it's been falsified by observation?" to which Kroupa replied, "Absolutely."[11]

In light of this development, the interviewer inquired as to why people cling onto the dark matter concept. Kroupa answered: "It has no rational basis....It's probably related to sociological pressures in the community....It's tribal thinking that pulls a lot of young people into an established [way of] thinking because that's where they can have a career, that's where the funding is, that's where the leaders of the field are famous. They get prizes, and that's a self-strengthening system in a scientific establishment which relies on competition for funding. That's the whole fallacy of the situation."[12]

To recap: Dark matter is a fundamental element of modern cosmology. It's needed in order to make the consensus model work, which includes preserving our current understanding of gravity and general relativity. And now scientists say that not only is dark matter mysterious, but it's actually been falsified and only remains propped up because of "sociological pressures." We've barely gotten started, and modern cosmology already appears to have extremely tenuous foundations.

If alleged "dark matter" isn't concerning enough, consider the story of alleged "dark energy." It was conceived in the late 1990s, although its origin traces back to 1912. A series of observations led physicists to conclude that the universe is expanding—and it's doing so more rapidly over time. This phenomenon is known as *cosmic acceleration*. As stated in a 2024 article published by the National Aeronautics and Space Administration (NASA): "Right now, dark energy is just the name that astronomers gave to the mysterious 'something' that is causing the universe to expand at an accelerated rate."[13]

The NASA article gives further information within the context of the mainstream cosmological story: "Some 13.8 billion years ago, the universe began with a rapid expansion we call the big bang. After this initial expansion, which lasted a fraction of a second, gravity started to slow the universe down. But the cosmos wouldn't stay this way. Nine billion years after the universe began, its expansion started to speed up, driven by an unknown force that scientists have named dark energy. But what exactly is dark energy? The short answer is: We don't know. But we do know that it exists, it's making the universe expand at an accelerating rate, and approximately 68.3 to 70% of the universe is dark energy."[14]

The 2006 "Report of the Dark Energy Task Force"—written by thirteen esteemed academics from Harvard, Princeton, MIT, and other institutions—gives further insights:

> **Dark energy appears to be the dominant component of the physical Universe, yet there is no persuasive theoretical explanation for its existence or magnitude.** The acceleration of the Universe is, along with dark matter, the observed phenomenon that most directly demonstrates that **our theories of fundamental particles and gravity are either incorrect or incomplete.** Most experts believe that nothing short of a revolution in our understanding of fundamental physics will be required to achieve a full understanding of the cosmic acceleration. For these reasons, **the nature of dark energy ranks among the very most compelling of all outstanding problems in physical science.**[15] [emphasis added]

But that's not all. Beyond the problems discussed with alleged dark matter and dark energy, physicists lack a unifying theory—known as a *theory of everything*. As stated by physicist Katherine Freese, a professor at the University of Texas at Austin: "What they're talking about is unifying all the forces of nature into a single one."[16]

More specifically, the struggle to find a theory of everything comes from the incompatibility of gravity/general relativity with quantum mechanics. General relativity often deals with large-scale gravitational effects and the bending of space-time, whereas quantum mechanics is typically characterized as dealing with the ultrasmall. Physicists argue that the theories work well on their own, but the equations metaphorically "blow up" when combined. Thus, scientists are desperately trying to find "quantum gravity," to no avail so far.[17]

To be clear: The two foundational theories of modern physics are not compatible within a unifying theory. On top of that, 96 percent of the universe is unexplained dark matter and dark energy. Our modern view of the cosmos is—objectively—upside down.

However, it can be tempting to excuse these shortcomings by saying, "There's a lot we don't know, and scientists will figure it out eventually. Just give them some time and money. Modern physics does a good enough job of predicting many things." This line of thinking carries with it a bias toward holding on to the consensus model of reality. And perhaps it's wishful thinking. **As stated by physicist Michio Kaku: "There is a crisis in cosmology.... Usually in science, if we're off by a factor of two or a factor of ten, we call that 'horrible.' We say something is wrong with the theory....However, in cosmology, we're off by a factor of 10^{120}. That is 1 with 120 zeros after it. This is the largest mismatch between theory and experiment in the history of science."[18]** [emphasis added]

An alternative thought process is needed. Perhaps something like this: *The reason 96 percent of the universe is unknown, and the reason we don't have a theory of everything, is that we have a fundamentally flawed way of looking at the cosmos and reality more broadly. Dark matter and dark energy don't actually exist. They aren't even "problems" that modern physics needs to "solve." Their alleged existence reflects a desperate attempt to preserve the current cosmological model, including gravity and general relativity theory. From this lens, the notion that dark matter and dark energy were even "discovered" is a misleading euphemism. It's perhaps more accurate to say that dark matter and dark energy were simply plugged in—they were invented out of thin air to avoid starting from scratch in our basic understanding of the cosmos. Then it makes sense why we don't have a theory of everything—because something is deeply wrong with our basic models. And until scientists reconsider their assumptions, they'll continue hitting a wall and misleading the public about the cosmos and Earth's place in it.*

The arguments in this book operate from the latter approach. It's a "start from scratch" mentality demanding that we question *everything* we think we know about the cosmos and Earth's place in it. *Everything.*

However, starting from scratch can be destabilizing. Thus, there's a psychological incentive to resist such an approach and cling on to the "old way" because it's familiar. The consensus-sustained cosmological model is also supported by revered scientific figures, despite its extreme shortcomings. There can be a tendency to struggle with the idea that intelligent scientists could be so wrong.

The quest for truth demands that we let go of such psychological barriers and look at the data objectively and rationally. Preconceived notions based on mere beliefs or arbitrary philosophical preferences need to be thrown out.

Why is this so important? Perhaps fundamental shifts in physics will lead to technologies that make the world a better place, with a drastically improved quality of life for all. Perhaps that will enable true freedom rather than dependence on external authorities. And perhaps reconsidering the cosmos, and our place in it, will reveal more about who we are, why we're here, what our true history is, and how we should be living our lives. The implications are ultimately massive.

In a 1995 interview, famed physicist Stephen Hawking summarized the implications of the mainstream cosmological model. He stated: "The human race is just a chemical scum on a moderate-sized planet, orbiting around a very average star in the outer suburb of one among a hundred billion galaxies. We are so insignificant that I can't believe the whole universe exists for our benefit."[19] But if Hawking's bleak remarks were based on faulty scientific assumptions, then what might be our *true* place in the cosmos?

In light of the foregoing, a deep questioning of the consensus cosmological model is urgently needed.

Challenging Thought Processes

Before diving into the specifics, first it's important to orient one's thinking appropriately. The subjects in this book are scientific in nature, but the hard part isn't as much the "science" as it is the thought processes that block rationality. What follows are cautions for you, the reader. Additionally, this will serve to prime you for paradigm-shifting concepts I discuss later that may cause cognitive dissonance (that is, psychological discomfort). Sometimes easing into jarring subject matter is most palatable to the mind.

1. *Falsifying one model does not require having a viable or complete alternative model.* When a claim is put forward, the burden of proof is on the party making the claim. If observations falsify the claim, then the claim must logically be discarded—regardless of whether a new, viable claim can be made to replace the old one.

   In other words, it's permissible to show that a model is incorrect without providing a new and comprehensive model to replace it. In such scenarios, we're left in the uncomfortable but intellectually honest position of saying, *We simply don't know the answer.*

   Cosmology researcher Austin Whitsitt describes this principle using the phrase **"falsification is independent of replacement"**: the exercise of falsifying one model is independent of finding a definitive replacement model.[20] This is a process of elimination that narrows down the possible truths.

   Here is an example: You're working with a team on a multiple-choice test in school that has options A, B, C, and D as possible answers. You're required to submit an answer to your teacher, but the caveat is that your entire team must agree on the answer that you submit. Your teammates are convinced that the answer is option A. However, your analysis leads you to conclude that A cannot be correct. You tell your team the news, excited that you narrowed down the set of possibilities. But to your shock, your team dismisses your work. Although they cannot deny that you've falsified option A, they ask with skepticism, *But do you know, instead, what the correct answer is? If it's not A, is it B, C, or D*? You reply, *I don't know. That's a separate analysis, and I haven't gotten there yet*. Your team replies, *You don't know the answer? Okay, then the real answer must be A. We're sticking with that*. This is irrational and illogical, but it can happen with rebuttals to scientific models.

Whitsitt gives another example that's illustrative: Imagine that you go to your parents' home and find in their closet official legal papers showing that you were adopted. All the paperwork is there: you were indeed legally adopted. Then you tell your friends, *Hey, you won't believe this: I was adopted! I found the paperwork!* Your friends reply, *Who are your real parents, then?* You reply, *I don't know. That requires a separate line of inquiry. I'd have to look into that much further, but the truth is that I might never know who my real parents are. All I know with certainty is that I was adopted*. Your friends reply, *We refuse to believe that you were adopted until you tell us who your real parents are*.[21] This is the same sort of irrationality that can occur within science: even after one model is falsified, there's a bias toward sticking with it if a comprehensive replacement model isn't provided.

The psychological desire to have complete answers clouds logical thinking, as opposed to being able to sit in the unknown. There is often a tendency to feel more comfortable living with a lie than living with uncertainty.

This concept is essential to hold in mind while reading the coming chapters. In this book, I make no definitive claims about the true nature of the cosmos—and for reasons that I will discuss, there are limitations that make it impossible to promote a definitive model today. Instead, in this book I aim to expose flaws in the claims put forward by mainstream science. Hopefully by deflating confidence in the existing model, more attention will go toward new and better models.

The approach is similar to the Vedic philosophy of *neti neti* ("not this, not that").[22] In other words, one arrives at the truth by continually finding what is *not* true, thereby leaving a narrower and narrower subset of possible candidates for the truth.

2. *Black swans falsify a model.* Imagine a model that claims: "All swans are white." You travel around the world, and you find white swan after white swan. You conclude that the model works well and is highly predictive of reality. Then, one day, you come across a black swan. That black swan destroys your model because your model says that "all swans are white."[23] *All* means *all.* A single anomaly makes your model unviable because your model is no longer able to accurately describe reality. This may sound obvious, but it needs to be stated explicitly. **There is often a psychological tendency to excuse mistakes and defend the consensus model by using mental gymnastics, even if contradictory evidence is presented.**

   These mental gymnastics often include post hoc rationalizations: an idea is made up out of thin air to "explain away" the apparent anomaly, thereby allowing that model to *seem like* it still works.

   Dark matter and dark energy might turn out to be examples of this rationalization. Observations in astrophysics failed to align with the mainstream model, so scientists effectively concluded: *Dark matter and dark energy must be there.* But what if they're *not* there, and the fundamentals of modern cosmology—including gravity and general relativity theory—are simply wrong?

3. *Double standards are everywhere in debates about cosmology.* If an existing model explains reality "pretty well," it's easy to sweep under the rug seemingly minor errors, incomplete ideas, or unexplained phenomena.

   However, when a new, competing model is proposed, there can be a tendency to quickly dismiss it if there is even a *single* unexplained issue…while at the same time being willing to tolerate unexplained issues in the existing, mainstream model. This is, of course, irrational and intellectually inconsistent, yet it happens regularly. If one is willing to be so harsh toward a new model, one should be equally ready to let go of the existing model.

For instance, mainstream science seems willing to accept the problems with dark matter and dark energy, and the absence of a unifying theory of physics. Yet when the "nonmainstream" ideas I discuss in this book are proposed, there's a knee-jerk reaction to dismiss them if there's even a single concept that is difficult to conceptualize.

4. *Observations do not, on their own, reveal the cause(s) of those observations.* Imagine a situation in which people get sick with similar symptoms after attending a party. Then they conclude that they "caught" something contagious. This is certainly one reasonable hypothesis. But to stop there would be illogical. The fact that people got sick after being in the same place at the same time doesn't necessarily mean that a contagious, microscopic germ caused the symptoms. People can get sick for many reasons that have nothing to do with germs. For instance, what if people at the party were exposed to the same toxin floating in the air, emanating from the materials in the building, or sitting in the food and drinks served at the party? What if the attendees were exposed to high levels of radiation or electric and magnetic fields (EMFs)? What if they were exposed to a similar psychological stressor that translated into sickness? What if there was a "bioresonance" between the "energy fields" of the people that mimicked germ-based contagion? Yawns seem to be "contagious," but they don't come from a germ. Women's menstrual cycles sometimes synchronize when they share a living space, but that isn't believed to come from a contagious germ. Furthermore,

what if there is some other cause that isn't understood by modern science?

Rather than really digging for the truth, it's easy to find ourselves in a place of intellectual laziness whereby a single explanation emerges, and we stop thinking any further. Once we've decided on the explanation, we erroneously use that as the basis for the model into which all future observations are forced.

This work is, of course, not about medicine (I covered that topic in my book *An End to Upside Down Medicine* [2023]). But the fallacious reasoning observed throughout medicine often occurs in cosmology as well. Andrew Kaufman, MD, made this very observation in his 2023 lecture titled "Fallacies, Fraud & Pseudoscience: The Parallels Between Cosmology and Biology." In his words: "You can observe a set of phenomena in nature and you can come up with a model or a hypothesis to explain it....But when you stick to circular reasoning, you feel that your explanation is the only possible solution....[It's advisable to] always think: 'If I can come up with another logical explanation for this set of phenomena, then we know that the proposed explanation or model is not necessarily correct because there are many other models that could also be correct.'"[24]

An example of this type of thinking arises in discussions about heliocentrism (the belief that Earth revolves around the Sun), versus geocentrism (the belief that the Sun revolves around Earth). Heliocentrism is fundamental to the consensus cosmological paradigm, yet physicists rarely acknowledge that observations attributed to heliocentrism *could also fit into a geocentric cosmology*. We're simply so accustomed to presupposing the truth of heliocentrism that we struggle to envision Earth at the center—*even though our experience is that the luminous bodies in the sky move around Earth, and Earth feels to us like it is stationary.* **Many prominent, mainstream physicists have discussed this, including**

Stephen Hawking. In his book *The Grand Design* (2010), coauthored by Leonard Mlodinow, they stated: "Our observations of the heavens can be explained by assuming either the earth or the sun to be at rest."[25] Let that sink in. There is more to come on this topic. [emphasis added]

The broader problem here is related to a logical fallacy known as *affirming the consequent*. It goes like this:

If P, then Q.

Q, therefore P.

An example:

If it rains, then the grass outside will be wet.

The grass outside is wet; therefore, it rained.

This conclusion, of course, ignores many other possible causes of wet grass that have absolutely nothing to do with rain. The grass could have been wet because of sprinklers, for instance.

How often do we get stuck in a particular belief system without considering other possible explanations?

5. *The way celestial bodies in the sky look and behave doesn't necessarily imply that Earth looks and behaves in the same way*. If you look at bodies in the sky through a telescope, it is not logical to conclude that Earth must share those same qualities. An alternative possibility—and one that must be considered among intellectually honest seekers—is that Earth is unique or distinct relative to those bodies observed in the sky. For instance, the assumption that Earth is just like "other planets" presupposes that Earth is a planet rather than something else. Alternatives need to be considered in fully comprehensive analyses.
6. *Testing something on Earth alone, in the present moment, is insufficient to draw conclusions about the broader cosmos in the past and in the future*. Scientists can run experiments on Earth today, but they often extrapolate and make

claims that ignore important presuppositions. As stated by physicist Robert Bennett, PhD:

> When we design a test, the test has to match what the hypothesis claims. **So, if you claim something that is going to apply throughout all space, you cannot test it only on Earth. If you say it's in the Solar System, you have to be able to test it throughout the Solar System.** This is what's overlooked very often in the statements of laws—the scope of the laws—in space.
>
> Same goes for time. Does the law only apply at a certain time? This is violated [frequently]....You cannot know what happened in ancient times if you make measurements on data in current time. You can only assume that nothing has changed between when the event occurred and when you actually did the measurement. And this is an assumption that is often incorrect.
>
> [Also,] when you look at the stars in space, what you're looking at are dots of light that are interpreted as being the same type of light production that is being done in the Sun. But there's no proof of that. The spectrum of the stars—none of them are the same as the spectrum of the Sun. So we're making assumptions that what we see in starlight is the same as we see in sunlight....
>
> [In summary,] when you look at a law, you must always ask, "What is the scope in space and time?" This is what I see most often violated in what I call "mainstream thinking."[26] [emphasis added]

7. *The way our eyes see obscures reality, and many of us are unaware of this issue; thus, we fail to make the necessary mental adjustments*. On the most fundamental level, our eyes only show us a tiny sliver of the electromagnetic spectrum: we don't see radio waves, x-rays, microwaves, infrared light, and more (see the image that follows). Even though we can detect invisible light with

technology, we tend to be biased by what we can see with our eyes. Therefore, judging the cosmos by our eyes alone is problematic.

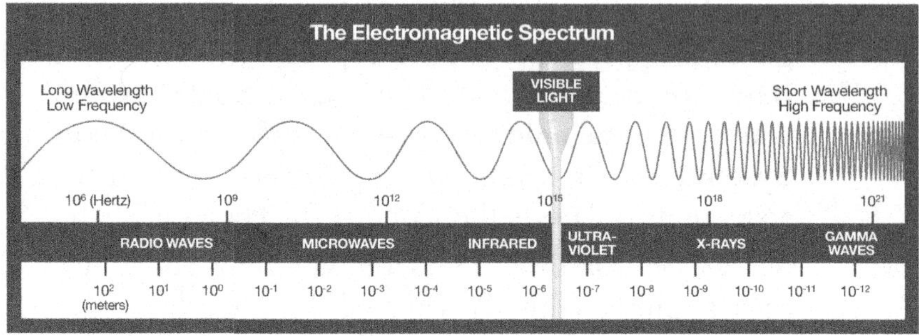

Visible light is only a tiny fraction of the electromagnetic spectrum. Most of the spectrum exists beyond what our ordinary vision can detect. Our eyes show us only a sliver of reality.[27]

Beyond that, it's important to acknowledge that we cannot see forever. Light attenuates, meaning that its intensity fades as it moves through a medium (such as the air and atmosphere).[28]

Similarly, a visual phenomenon known as *perspective* can lead us to erroneously interpret objects to be rising or falling as they become more distant from us, when in reality they're not actually rising or falling. Put another way: our brains process visual information in a way that can skew perception. Ultimately, our visual space is curved and limited. It is neither linear nor infinite. More on this later.

8. *Math is a tool; it is not direct evidence that a model is correct.* Consider the following quotation from Arthur C. Clarke in his 1963 book *Profiles of the Future*:

> The lesson to be learned...is one that can never be repeated too often, and is one that is seldom understood by laymen—who have an almost superstitious awe of mathematics. But mathematics is only a tool, though an immensely powerful one. **No equations,**

> **however impressive and complex, can arrive at the truth if the initial assumptions are incorrect.** It is really quite amazing by what margins competent but conservative scientists and engineers can miss the mark, when they start with the preconceived idea that what they are investigating is impossible. When this happens, the most well-informed men become blinded by their prejudices and are unable to see what lies directly ahead of them. What is even more incredible, they refuse to learn from experience; they will continue to make the same mistake over and over again.
>
> Some of my best friends are astronomers, and I am sorry to keep throwing stones at them—but they do seem to have an appalling record as prophets."[29] [emphasis added]

Physicist Nikola Tesla stated it another way in 1934: "Today's scientists have substituted mathematics for experiments, and they wander off through equation after equation, and eventually build a structure which has no relation to reality."[30]

In this book, I will not delve into math. But inevitably the question may arise in your mind: *Wouldn't equations have proven modern cosmology already?* If the assumptions baked into the equations haven't been correct, however, then the resulting math wouldn't tell us what we need to know. And perhaps that's what has been happening in cosmology.

9. *Beware of the "appeal to personal incredulity" logical fallacy.* When encountering paradigm-shifting concepts, there can be a tendency to think, *I cannot conceive how this could possibly be true. I cannot, in my own mind, list all of the steps needed in order for it to have happened in the way that it is claimed. I don't fully understand it. Therefore, I reject it.*

> Truth doesn't depend on our ability to fathom it. Just because one can't comprehend something doesn't mean that this thing is not true. Similarly, there might be ideas beyond our current knowledge that we haven't even considered—"unknown unknowns."[31]

What's to Come in This Book

Keeping in mind the discussed psychological challenges, part I explores our place in the cosmos. Chapter 1 examines the Sun-centered (heliocentric) model and its implications. Next, I discuss the Big Bang theory of our origin (chapter 2), followed by explorations about how to determine whether Earth is moving or stationary. This includes challenges to Einstein's relativity theory and Newton's theory of gravity (chapter 3).

Part II is a discussion on the nature of where we live. I present alternative ideas about Earth's shape in chapter 4, followed by phenomena that seemingly challenge the consensus "globe" model (chapter 5). In chapter 6, I delve into concerns about NASA's history, the Moon-landing missions, and data obtained by space agencies in general.

Although the discussions up to this point focus on the *physical* nature of the cosmos, in part III, I examine metaphysics. In chapter 7, I investigate a metaphysical framework centered around consciousness; and in chapter 8, I discuss implications within the context of an apparent "spiritual war" between dark and light metaphysical forces.

I conclude the book with a discussion about the importance of shifting our consciousness.

As another reminder, throughout this work, I do not make definitive claims about what is true, but rather seek to find out what is perhaps *not* true. Thus, discussions of alternative theories should be taken as suggestions for further exploration rather than assertions that they are accurate. All ideas I present here are intended to stimulate creative thinking, further research, a more liberated mind, and ultimately...a more liberated world.

PART I
WHAT IS OUR PLACE IN THE COSMOS?

CHAPTER 1

SUN-CENTERED HISTORY AND PHILOSOPHY

Given the fundamental problems of modern cosmology, its basic tenets need to be reexamined. One such tenet is heliocentrism, which is the idea that Earth revolves around the Sun. And furthermore, in this model, the Sun is at the center of the solar system. The contrasting framework—geocentrism—asserts that Earth is at the center and the Sun revolves around it. Modern cosmology's heliocentric bias informs how we think—not just about Earth, but ultimately about the broader cosmos, its origins, and our place in it.

However, as mentioned in the introduction via Stephen Hawking and Leonard Mlodinow's admission, *either* heliocentrism or geocentrism *could* be used to explain our observations of the sky. Other prominent thinkers have said the same thing.

For example, astronomer Fred Hoyle stated in 1973: "Let it be understood at the outset that it makes no difference, from the point of view of describing planetary motion, whether we take the Earth or the Sun as the center of the solar system. Since the issue is one of relative motion only, there are infinitely many exactly equivalent descriptions referred to different centers—in principle any point will do, the Moon, Jupiter."[1] Similarly,

physicist I. Bernard Cohen wrote in his book *The Birth of a New Physics* (revised and updated in 1985): "There is no planetary observation by which we on earth can prove that the earth is moving in an orbit around the sun."[2] [emphasis added]

So, if heliocentrism and geocentrism could be regarded as "equivalent," in a sense, why has heliocentrism been adopted as the default truth about the cosmos? And what are the philosophical implications of heliocentrism? In this chapter, I delve into these questions.

A Brief History of Heliocentrism and Geocentrism

The debate over heliocentrism versus geocentrism has a long history. Ancient Greeks such as Pythagoras (6th-century BCE) and Aristotle (4th-century BCE) held geocentric views based on their overall philosophy about Earth's place in the cosmos.[3] The most famous geocentrist was the Egyptian mathematician Ptolemy (around 150 CE). His views were described in his books *Almagest* and *Planetary Hypotheses*, and he was able to make many accurate predictions about the motions of the bodies in the sky using geocentrism.[4]

Heliocentrism has a long history as well, and it was promoted by the Greek astronomer and mathematician Aristarchus around the 3rd-century BCE.[5] In the 16th-century CE, Nicholas Copernicus's heliocentric model famously led to its predominance in modern thinking.[6] However, as noted in 2011 by physicist Carlo Rovelli: "Copernicus's predictive system is less accurate, not more, than Ptolemy's [geocentric system]."[7] Yet Copernicus's heliocentric model led to a scientific revolution—a paradigm shift—that still impacts our thinking today.[8]

As noted by the *Stanford Encyclopedia of Philosophy*, "It is impossible to know exactly why Copernicus began to espouse the heliocentric cosmology. Despite his importance in the history of philosophy, there is a paucity of primary sources on Copernicus....Many of the answers to the most interesting questions about Copernicus's ideas and works have been the result of conjecture and inference, and we can only guess why Copernicus adopted the heliocentric system."[9] Copernicus died shortly after the publication of his famous book, *On the Revolutions of the Heavenly Spheres*, in 1543.

Copernicus's heliocentrism was a controversial idea. It countered the geocentric cosmology that supported the Church's religious view—which was that God created Earth and that Earth was *at the center* of God's creation. As a result, Copernicus's book was banned in 1616. But, as noted in a 2016 article published by Jessica Wolf at UCLA: "After a few minor edits, making sure that the [heliocentric] theory was presented as purely hypothetical, [Copernicus's book] was allowed again in 1620 with the blessing of the church."[10]

Moreover, in 1616—the same year that Copernicus's book was initially banned—Italian astronomer Galileo Galilei was warned by the authorities not to endorse the Copernican perspective. He did it anyway, allegedly because of observations he made through his telescope. And he's now considered a hero for promoting the heliocentric revolution.

In 1633, after facing an inquisition, Galileo was convicted of a "strong suspicion of heresy" (a less severe conviction than one of true heresy). But perhaps Galileo's punishment was not as dramatic as it's sometimes depicted. UCLA professor Henry Kelly, who thoroughly examined the Church's investigation into Galileo, comments on this further: "In sum, the 1616 event was not the beginning of a 17-year-long trial, as is often said, but a non-trial....Galileo's actual trial lasted for only a fraction of a single day, with no fanfare at all." Kelly adds that Galileo was not tortured, but rather was *threatened* with torture.[11] Galileo's book was banned, he spent a day in prison, and then he was commuted to comfortable "villa arrest" in his Florence home for the rest of his life until his death in 1642.[12]

Brother Guy J. Consolmagno, PhD, an MIT– and University of Arizona–trained astronomer, and director of the Vatican Observatory, further commented on Galileo's situation in 2004:

> Nobody knows really why Galileo was gone after....For most of Galileo's life he was lionized, he was treated like a hero, including by people in the Church....When Galileo got into trouble at the end of his life, it was a real shock. It was a complete reversal of everything that had been said up to that point.
>
> And so the historical question is, why did it happen? And the answer is, we don't know. You can go to amazon.com and find 300 books on Galileo, every one of them with a different answer. Which is to say, there was something going on, and it wasn't simply a science versus religion thing....If you relied on *JFK* the movie to figure out what happened in the assassination of Kennedy, you'd be in as good shape. You've got to remember the Galileo affair occurred at the height of the Reformation and the 30 Years' War.[13]

UCLA's Kelly summarizes the situation well: "We can only guess at what [Galileo] really believed."[14] And yet today, Galileo is typically lauded as a prominent figure in the shift to heliocentrism.

Furthermore, author and filmmaker Robert Sungenis, PhD, adds a detail to the story that also challenges the traditional narrative:

> Few are aware that a year before he died Galileo renounced, quite dramatically, all his claims that the Earth went around the sun. In 1641 his colleague Francesco Rinuccini had written a letter to Galileo claiming that the astronomer Giovanni Pieroni, by discovering a [phenomenon of the stars], had proven that the Earth moves. Galileo, who by this time had had a dramatic conversion back to the Christian faith two years earlier in 1639, wrote [a] letter back to Rinuccini later in 1641 [seemingly withdrawing his support of the Copernican view]....
>
> Rinuccini was so disturbed by the letter that he attempted to erase Galileo's signature from off the last page. The editor of Galileo's works states: "The

> signature 'Galileo Galilei' has been very deliberately and repeatedly rubbed over, with the manifest intention of rendering it illegible." Stillman Drake, one of the premier Galileo historians of our day, states: "Among all Galileo's surviving letters, it is only this one on which his name at the end was scratched out heavily in ink. I presume that Rinuccini valued and preserved Galileo's letters no matter what they said, but did not want others to see this declaration by Galileo that the Copernican system was false, lest he be thought a hypocrite."[15]

So perhaps the "heliocentric revolution" wasn't quite as solid as it's often depicted. Yet, regardless of Galileo's true stance, his impact has undoubtedly been immense. The observations he made through his telescope shifted societal thinking around our place in the cosmos.

For instance, he observed the phases of the planet Venus (which can be likened to phases of the Moon). He reasoned that the phases of Venus would be explained easily if Venus orbited the Sun, suggesting that the Sun was at the center of the solar system. Galileo also saw moons circling around Jupiter. As stated in a NASA article, this observation provided "strong evidence for the Copernican theory that most celestial objects did not revolve around the Earth."[16]

In reality, however, heliocentrism wasn't the only way to explain Galileo's observations. Venus's phases could be explained with a geocentric model if the planets had an orbital relationship with the Sun while, at the same time, they *all* revolved around Earth. In other words, both heliocentrism and geocentrism could explain Galileo's observations. But the heliocentric explanation was the one that stuck.

Similarly, Galileo's observation that moons orbited Jupiter was not exclusive proof of heliocentrism, either. Another possibility was that Jupiter and its moons all revolved around Earth. Moreover, it is generally inappropriate to conclude that because smaller bodies (moons) revolved around a larger body (Jupiter), Earth similarly revolves around the Sun. That's certainly *one* possibility, but not the only one. Fundamentally, it's important to keep in mind that observing Jupiter and its moons...is observing *Jupiter and its moons*. That's not the same thing as observing Earth. It would be illogical, and simplistic, to conclude that because Jupiter and its moons behave and look a certain way, Earth shares Jupiter's qualities. This is akin to concluding that we know about the floor by studying the ceiling.[17] As written by Sungenis in his 2017 book *Geocentrism 101*: "What [Galileo] claimed as proof then would be summarily dismissed out of science classrooms today. Galileo merely began a hypothesis—a hypothesis that would eventually take on a life of its own and become the status quo of popular thinking."[18]

The next major historical milestone was the emergence of Tycho Brahe's work (1546–1601). His model was neither purely Ptolemaic nor Copernican, but rather a mixture of the two. It was *somewhat* "heliocentric," in that it argued that planets revolved around the Sun, but fundamentally geocentric in that the Sun revolved around Earth. This is the sort of cosmology that could explain Galileo's observations of Venus's phases.[19]

Brahe's assistant, Johannes Kepler (1571–1630), took over Brahe's treasure trove of astronomical data following his death. But rather than furthering the geocentric hypothesis, Kepler fit Brahe's data into the heliocentric model and refined what Copernicus had put forth.[20]

However, some speculate that foul play was involved—in order to promote heliocentrism. A leading theory is that Kepler murdered Brahe in order to get ahold of Brahe's records. As written in the 2004 book *Heavenly Intrigue*, Joshua Gilder and Ann-Lee Gilder contend: "For four hundred years it was believed that Tycho Brahe died of natural causes....Recent forensic analysis of remnants of his hair reveals that he was murdered, systematically poisoned [with mercury]. And all the answers to the motive, means, and opportunity point directly to one suspect: Johannes Kepler."[21] They also observe: "If it hadn't been for Tycho Brahe, Kepler would be a mere footnote in today's science books."[22] In 2010, Brahe's body was exhumed, for the second time, and according to a BBC article, the levels of mercury found were "not high enough to have killed him."[23] What actually happened remains a topic of debate.

Regardless of whether Kepler murdered Brahe, what's not debated is the impact of his work. In fact, Kepler's heliocentric work influenced another famous heliocentrist—and originator of the theory of gravity—Isaac Newton (1643–1727).[24] In Newton's view, the revolutions of celestial bodies were actually revolutions around a center of mass. Because the Sun had so much more mass than Earth, Earth revolved around the Sun. This also aligned with Galileo's observations of Jupiter's moons orbiting Jupiter (smaller bodies revolving around a bigger one).[25]

However, we're not often told that Newton acknowledged other possibilities. **Physicist Steven Weinberg writes in his 2015 book, *To Explain the World*: "In an unpublished 'Proposition 43' that did not make it into [Newton's famous writing] *The Principia*, Newton acknowledges that Tycho's [geocentric] theory could be true if some other force besides ordinary gravitation acted on the Sun and planets."[26]** [emphasis added]

Contemporary physicists such as André Assis (from Brazil) have also stated that a geocentric model could be true. He wrote in 1999 that under relational mechanics, "the earth can remain at rest and at an essentially constant distance from the sun."[27] Physicist Luka Popov made similar remarks in his 2013 paper, "Newtonian–Machian analysis of the neo-Tychonian model of planetary motions," published in the *European Journal of Physics*.[28]

So perhaps the "Copernican Revolution" isn't as solid as we're all told, and geocentric possibilities need to be taken seriously. And, to reiterate, many prominent physicists have acknowledged this vital idea. Astronomer Fred Hoyle summed up the situation well in his 1973 book, *Nicolaus Copernicus: An Essay on his Life and Work*: "Today we cannot say that the Copernican [heliocentric] theory is 'right' and the Ptolemaic [geocentric] theory 'wrong' in any meaningful physical sense. The two theories...are physically equivalent to one another."[29] Assis expressed it in another way:

"We have found a complete equivalence between the Ptolemaic and Copernican world systems. It is then equally valid to say that the earth moves relative to the distant universe, or that it is at rest and that it is the distant universe which moves relative to the earth."[30]

Even Albert Einstein and his coauthor Leopold Infeld remarked in their 1938 book, *The Evolution of Physics*: "The struggle, so violent in the early days of science, between the views of Ptolemy and Copernicus would then be quite meaningless. Either [coordinate system] could be used with equal justification. The two sentences, 'the Sun is at rest and the Earth moves,' or 'the Sun moves and the Earth is at rest,' would simply mean two different conventions concerning two different [coordinate systems]."[31]

The Copernican Principle

If the heliocentric and geocentric models are as equivalent as many respected physicists have confessed...then why is heliocentrism the default modern worldview? It's so deeply embedded in everyday thinking that most of us probably haven't thought to question it.

The 2022 documentary *Heliosorcery: Exposing the Occult Origins of Heliocentrism* makes the case that the push for heliocentrism is part of a long-standing conspiracy by sun worshippers to misdirect the public, ultimately making way for Darwin's theory of evolution.[32] The documentary even refers to the Galileo affair with the Church as a "publicity stunt" and further implicates Copernicus, Kepler, and Newton.[33]

This matter is approached a bit differently in a 2014 documentary produced by Robert Sungenis called *The Principle*.[34] **The documentary exposes the bias in mainstream science toward the belief that "we are not special"—known as "The Copernican Principle."** As summarized by University of Washington astrophysicist Jonathan I. Katz in 2002:

> In school we are taught that...Nicolaus Copernicus, established that Earth and the planets revolve around

> the sun, rather than the planets and sun revolving around earth....The importance of Copernicus's ideas was both philosophic and scientific: **Man is not at the center of the universe, but is only an insignificant spectator, viewing its fireworks from somewhere in the bleachers.** In modern times this has been elevated into the "cosmological principle," which states that, if averaged over a sufficiently large region, the properties of the universe are the same everywhere; our neighborhood is completely ordinary and unremarkable. **We are not special, and our home is not special, either. This is one of the foundations of nearly all modern cosmologies.**[35] [emphasis added]

Conversely, if geocentrism is true and Earth *is* at the center, then perhaps we do occupy a special place in the cosmos. And perhaps there is a higher intelligence that made it happen. Such a perspective is compatible with religious and spiritual perspectives. On the other hand, heliocentrism and the Copernican Principle open the door for nihilistic atheism much more easily. If we don't occupy a special place in the universe, perhaps our existence is random, meaningless, and godless.

The modern scientific establishment tends to view religious and spiritual perspectives as superstitions of the past and thus tends to be biased against geocentrism. Consequently, mainstream scientific thinkers often demonstrate a *philosophical preference* for heliocentrism, which is, of course, not a *scientific reason* to believe something. Stephen Hawking exemplified this perspective in *A Brief History of Time* (1988) when he said: "We believe it only on grounds of modesty."[36]

The Copernican Principle is indeed "modest" in that it doesn't regard us as occupying a special place in the cosmos. "Modesty" alone shouldn't be a sufficient foundation on which to build an entire cosmology. And yet that's exactly what seems to have happened.

George Ellis, professor emeritus at the University of Cape Town in South Africa, reiterated this sentiment in 1995: "People need to be aware that there is a range of models that could explain the observations. For instance, I can construct [for] you a spherically symmetrical universe with Earth at its center, and you cannot disprove it based on observations. **You can only exclude it on philosophical grounds.... What I want to bring into the open is the fact that we are using philosophical criteria in choosing our models. A lot of cosmology tries to hide that."**[37] [emphasis added]

A "modest" Copernican philosophy also has implications for the apparent fine-tuning of the cosmos that enables human life. As Neil deGrasse Tyson puts it: "Earth formed in a kind of Goldilocks zone around the Sun, where oceans remain largely in liquid form. Had Earth been closer to the Sun, the oceans would have evaporated. Had Earth been much farther away, the oceans would have frozen. In either case, life as we know it would not have evolved."[38]

Few seem to debate whether the cosmos is perfectly aligned for human life to exist, but many *do* debate what caused this to happen. Is the fine-tuning of the cosmos evidence of a higher intelligence, or is it simply a coincidence? The implications of the "modest" Copernican Principle naturally steer scientists toward the latter perspective. Physicist Lawrence Krauss expresses this view directly: "The reason the universe appears to be so tuned for our existence is not because some Divine intelligence decided, *I want to create a universe so people can be in it*, but rather, if it were any different, we wouldn't be here."[39]

CHAPTER 2

HOW THE COSMOS BEGAN

The Copernican Principle—which posits that we do not occupy a special place in the cosmos—has deeply impacted scientists' views about the universe's origin. In fact, this bias might have led to inaccurate cosmological stories that could be better explained with a geocentric model. This chapter examines mainstream science's approach to these matters and discusses why the consensus model deserves to be challenged.

Edwin Hubble and the Expanding Universe

In the 1920s, American astronomer Edwin Hubble (1889–1953) used the 100-inch Hooker Telescope at Mount Wilson Observatory to observe galaxies that he considered to be outside of the Milky Way. He largely found that their light exhibited a "red shift," which he interpreted as evidence that the galaxies were moving farther and farther away from Earth. This suggested to him that we are in an expanding universe. Then he reasoned that there must have been a beginning point to the universe, which later led to the Big Bang theory.[1]

Stephen Hawking commented on the significance of Hubble's unexpected findings: "At that time most people expected the galaxies to be moving around quite randomly, and so expected to find as many blue shifted spectra as red shifted ones. It was quite a surprise, therefore, to find that most galaxies appeared red-shifted: nearly all were moving away from us!"[2] One interpretation of this "surprise" data could have been that Earth was at the center of the cosmos—and that's why everything looked like it was moving away from us.[3] As Hawking further remarked: "Now at first sight, all this evidence that the universe looks the same whichever direction we look might seem to suggest there is something special about our place in the universe. **In particular, it might seem that if we observe all other galaxies to be moving away from us, then we must be at the center of the universe.**"[4] [emphasis added]

But instead of advancing this geocentric interpretation, Hubble ultimately demonstrated a philosophical preference toward the Copernican Principle. His bias continued to move science away from the religious or spiritual notion of a "higher intelligence" that was involved in our creation.

Hubble even stated in his 1937 book, *The Observational Approach to Cosmology*, that he preferred not to choose an interpretation suggestive of geocentrism: **"Such a condition would imply that we occupy a unique position in the universe, analogous, in a sense, to the ancient conception of a central earth [that is, geocentrism]. The hypothesis cannot be disproved but it is unwelcome and would be accepted only as a last resort in order to save the phenomena. Therefore, we disregard this possibility and consider the alternative.... The unwelcome supposition of a favored location must be avoided at all costs."**[5] [emphasis added]

A model that "avoided" geocentrism "at all costs" involved constructing a universe that had no center. Hubble remarked on what this would look like: "All observers, regardless of their location, will see the same general picture of the universe.... There must be no favored location in the universe, no center, no boundary; all must see the universe alike. And, in order to ensure this situation, the cosmologist postulates spatial isotropy and spatial homogeneity, which is his way of stating that the universe must be pretty much alike everywhere and in all directions."[6] (Note: *Isotropy* means the universe will look the same in all directions, and *homogeneity* means that the universe should have the same basic structure in all places. Remember this phrase, "isotropy and homogeneity," because, as we'll soon see, it presents a big problem for more recent cosmological findings.)

Hubble further commented that he didn't want to see Earth with a special position in the cosmos: "A favored position, of course, is intolerable; moreover, it represents a discrepancy with the theory, because the theory postulates homogeneity. Therefore, in order to restore homogeneity, and to escape the horror of a unique position, the departures from uniformity, which are introduced by the recession factors, must be compensated by the second term representing effects of spatial curvature. There seems to be no other escape."[7] [emphasis added]

Hubble's Copernican bias was obvious. The implications of geocentrism were "intolerable" to him, and he felt as though there needed to be an "escape" from the "horror" of that possibility.

Robert Sungenis summarizes the situation concisely: "The 'Copernican' bias drove modern cosmology to its presently accepted interpretation of Hubble's data: instead of galaxies being organized around a common center as the evidence plainly showed, they insisted, rather, that the universe is merely a balloon-like surface wherein space is curved. The galaxies are said to lie on this curved surface and are spreading out from each other; and most important, Earth is in one of those moving galaxies, not in the center of the universe, for there is no center to a balloon universe."[8] What follows is an illustration of the expanding universe using a balloon analogy.

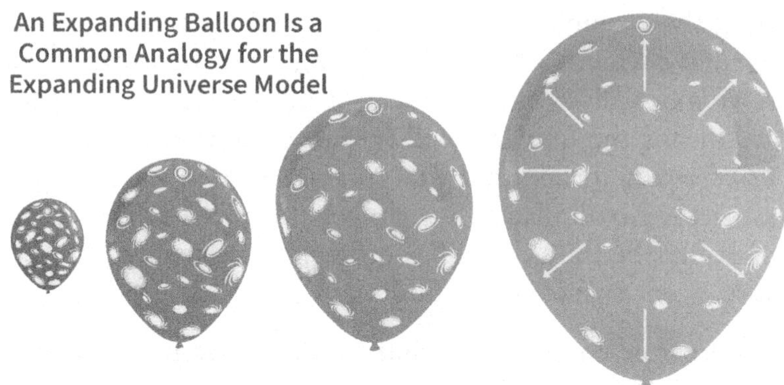

The expanding universe model is generally likened to a balloon. Imagine a balloon with dots drawn on it. When air is blown into the balloon, the surface expands such that the dots are farther and farther away from one another. This basic analogy explains what happens to celestial bodies as the universe itself is expanding —at least as it's posited by mainstream cosmologists. However, as discussed, there are many reasons to question this model.

In light of all of this, one might wonder: *Is mainstream science's origin story of the cosmos a mere creation of the human mind, designed to ensure that we don't think we occupy a special place?*

The Big Bang

In the late 1920s, Belgian priest and scientist Fr. Georges Lemaître created a more formal cosmological model now known as the *Big Bang*, which involved a "cosmic egg" that exploded. This led to the expansion of the universe and set the stage for the formation of Earth.[9]

But Lemaître didn't call it a Big Bang. The term itself was coined by physicist Fred Hoyle in 1949 during radio lectures.[10] Ironically, Hoyle never truly embraced the theory himself. In the subsequent decades, he was discouraged because the Big Bang wasn't able to explain various phenomena. He remarked: "I began to get the sense that there was something seriously wrong…I'm a great believer that if you have a correct theory, you show a lot of positive results. It seems to me that they'd gone on for 20 years, by 1985, and there wasn't much to show for it. And that couldn't be the case if it was right."[11]

An ongoing problem for the Big Bang theory is known as the "Hubble tension." As NASA describes it: "The universe is now expanding faster than predicted based on observations of how it looked shortly after the big bang....There may be something woven into the fabric of space that we don't yet understand."[12] NASA leaves out an obvious possibility to explain the unresolved Hubble tension: perhaps something is fundamentally wrong with the Big Bang theory itself. **As in: maybe there was no Big Bang as it's currently conceived.** Therefore, some cosmologists even refer to the Hubble tension as a "crisis" for modern cosmology.[13]

But the Hubble tension isn't the only obstacle for the Big Bang theory. NASA lists several other persistent problems:[14]

- *The Flatness Problem:* The overall geometry of the universe is nearly flat, whereas Big Bang cosmology would predict a growing curvature. As NASA puts it: "A universe as flat as we see it today would require an extreme fine-tuning of conditions in the past, which would be an unbelievable coincidence." NASA reflects a Copernican bias here by not wanting to attribute the fine-tuning of the cosmos to anything other than "an unbelievable coincidence."
- *The Horizon Problem:* Distant regions of space that should never have been in contact according to Big Bang cosmology seem to have been in contact in the past.
- *The Monopole Problem:* Big Bang cosmology predicts that "magnetic monopoles" should have been created in the early universe, but they have not been found.

Rather than reconsidering the Big Bang theory itself, physicists have attempted to accommodate these problems by positing a rapid, strong inflation in the early universe around the time of the Big Bang. They've even invented hypothetical "inflaton" particles that inflate space, but no such particles have been found by physicists.[15] (Note: This is reminiscent of "dark matter" and "dark energy," as discussed in the introduction. Perhaps such entities, including hypothetical inflatons, are all merely post hoc rationalizations designed to preserve a cosmology centered around the Big Bang and heliocentrism. All such alleged entities—inflatons, dark matter, and dark energy—are unknown, unestablished, and abstract, but they are often spoken about as if they have already been established. This is known as the *reification fallacy*—speaking about an abstraction as if it's real, which serves to reinforce the belief in its existence.)

The Cosmic Microwave Background

Another major problem for Big Bang cosmology, particularly as it pertains to the Copernican Principle, comes from the "cosmic microwave background" (CMB). The CMB refers to a predicted background radiation that should pervade the universe—a remnant of the immense heat of the early universe following the Big Bang. The CMB was initially predicted in 1948. It was, in fact, later observed, which *seemed* like a major victory for the Big Bang theory. NASA summarizes this important history:

> [The CMB] was first observed inadvertently in 1965 by Arno Penzias and Robert Wilson at the Bell Telephone Laboratories in Murray Hill, New Jersey. The radiation was acting as a source of excess noise in a radio receiver they were building. Coincidentally, researchers at nearby Princeton University, led by Robert Dicke and including Dave Wilkinson...were devising an experiment to find the CMB. When they heard about the Bell Labs result they immediately realized that the CMB had been found. The result was a pair of papers in the *Astrophysical Journal* (vol. 142 of 1965): one by Penzias and Wilson detailing the observations, and one by Dicke, Peebles, Roll, and Wilkinson giving the cosmological interpretation. Penzias and Wilson shared the 1978 Nobel prize in physics for their discovery.[16]

After this preliminary investigation into the CMB, things were looking pretty good in terms of validating the Big Bang. But recall that according to the expanding-universe cosmology that Hubble helped to conceive, the universe should be isotropic (no preferred direction) and homogenous (uniform).

In contrast to these predictions, further investigations of the CMB have consistently found that it was neither isotropic nor homogenous (it was *an*isotropic and *in*homogeneous). This was first uncovered by NASA's Cosmic Background Explorer (COBE, 1989–1993), again by NASA's more precise Wilkinson Microwave Anisotropy Probe (WMAP, 2001–2010), and then again by the even more detailed Planck mission (2009–2013), which was operated by the European Space Agency (ESA) and received contributions from NASA. These missions keep looking, and they keep finding anomalies.

The ESA noted about the Planck mission: "One of the most surprising findings is that the fluctuations in the cosmic microwave radiation (CMB) temperatures at large angular scales **do not match those predicted by the standard [Big Bang] model**.... Another is an asymmetry in the average temperatures on opposite hemispheres of the sky. **This runs counter to the prediction made by the standard [Big Bang] model that the Universe should be broadly similar in any direction we look.**"[17] The ESA further remarked: "We see an almost perfect fit to the standard [Big Bang] model of cosmology, **but with intriguing features that force us to rethink some of our basic assumptions.**"[18] [emphasis added]

Robert Sungenis notes the "contradictory" aspects of this claim—the data was somehow "almost perfect" and yet required rethinking of "basic assumptions." The first part of the statement sugarcoats how problematic the CMB data is for the Big Bang cosmology. Sungenis further discusses the implications from his perspective: "In essence this means the Big Bang theory and the cosmological principle were falsified by the 2013 Planck [CMB] data. But this is being finessed and glossed over due to a prior commitment to the Copernican Principle."[19]

Perhaps the Big Bang theory is not as solid as we've been told.

Moreover, the CMB data has been interpreted to question Earth's place in the cosmos. Even physicist Lawrence Krauss, a popular advocate of mainstream cosmology, commented in 2006 on the potentially geocentric implications of the CMB data: "When you look at [the] CMB map, you also see that the structure that is observed, is in fact, in a weird way, correlated with the plane of the earth around the sun. Is this Copernicus coming back to haunt us? That's crazy. We're looking out at the whole universe. There's no way there should be a correlation of structure with our motion of the earth around the sun—the plane of the earth around the sun—the ecliptic. **That would say we are truly the center of the universe."**[20] [emphasis added]

In fact, other physicists have interpreted the CMB data to be in alignment with Earth, including its equinoxes. Could this mean that geocentrism is true? This is a perplexing concept for physicists operating within the Copernican Principle. For instance, University of Michigan researchers commented in 2004: "Physical correlation of the cosmic microwave radiation (CMB) with the equinoxes is difficult to imagine, since the WMAP satellite has no knowledge of the inclination of the Earth's spin axis."[21]

Furthermore, astrophysicist Rahul Kothari and his colleagues noted in 2013 that the data "suggests a potential violation of the cosmological principle" (recall that the cosmological principle is an extension of the Copernican Principle).[22] In Ashok Singal's 2013 paper titled "Is there a violation of the Copernican Principle in radio sky?," he similarly remarks: "There is certainly cause for worry.... **The Copernican Principle seems to be in jeopardy.**"[23] [emphasis added]

To Sungenis, an avid geocentrist, the accumulated CMB data suggests the following: "If we were forced to pick a place in which the Earth resided based on the equatorial and ecliptic alignment of the [CMB's] axes, the Earth would be at the very point of intersection."[24] In other words, from this lens, the CMB data could be interpreted to suggest that the cosmos is geocentric.

But that is an "unwelcome" notion to mainstream cosmologists (as Hubble phrased it). In fact, starting in 2005, the intersection described above by Sungenis became known as the **"Axis of Evil."** This term was adapted from the George W. Bush administration's geopolitical "Axis of Evil," which included Iraq, Iran, and North Korea.

It is noteworthy that a scientific finding would be named using the word *evil*. That implies a moral judgment. **The term *Axis of Evil* illustrates the scientific establishment's allegiance to the heliocentric model. Heliocentrism is almost like a scientific "religion" that isn't supposed to be challenged—which is ironic because the Copernican Principle often thrives because of the rejection of religion.**

But some physicists *have* acknowledged the profound implications. For example, Brazilian physicist Joao Magueijo told *New Scientist* in 2005 that the Axis of Evil "could be telling us something fundamental about our universe, maybe even that the simplest big bang model is wrong."[25] He also alluded to the possibility that the data could be explained by "a rotating universe."[26] It's reasonable to at least ask: Might the universe rotate around *Earth*? Is that what the CMB data is telling us?

Galactic Challenges to Heliocentrism

Observations of quasars (bright celestial bodies) have also challenged the Copernican Principle. Physicist Y. P. Varshni discussed this in a 1976 paper published in *Astrophysics and Space Science*: "Assuming the cosmological red shift hypothesis, the quasars… are arranged on 57 spherical shells **with Earth in the center**. This is certainly an extraordinary result. Some of the possibilities that we shall consider to accommodate this result may be disturbing, but we must consider these possibilities dispassionately." His use of the word *disturbing* indicates an acknowledgment of the prevailing heliocentric bias. Varshni continued: "**The Earth is indeed the center of the Universe. The arrangement of quasars on certain spherical shells is only with respect to the Earth. These shells would disappear if viewed from another galaxy or**

quasar. This means that the cosmological principle will have to go. Also it implies that a coordinate system fixed to the Earth will be a preferred frame of reference in the Universe. Consequently, both the special and general theory of relativity must be abandoned for cosmological purposes."[27] Here, Varshni goes a step further than simply challenging heliocentrism; he's even challenging Einstein. Varshni's results were validated in a 2001 paper by Ukrainian astronomers and were reinforced by University of Michigan physicist Michael Longo.[28] [emphasis added]

Beyond Varshni's work on quasars, research published in 2016 by physicists Behnam Javanmardi and Pavel Kroupa is suggestive of *galaxy* alignments with Earth (see their article "Anisotropy in the all-sky distribution of galaxy morphological types"[29]).

And physicist Jonathan Katz observed gamma ray bursts—high-energy light—that further challenge the Copernican Principle. He wrote in 2002:

> The uniform distribution of burst arrival directions tells us that the distribution of gamma-ray-burst sources in space is a sphere or spherical shell, **with us at the center** (some other extremely contrived and implausible distributions are also possible). But Copernicus taught us that we are not in a special preferred position in the universe; Earth is not at the center of the solar system, the sun is not at the center of the galaxy, and so forth.... This is the Copernican dilemma....
>
> **No longer could astronomers hope that the Copernican dilemma would disappear with improved data. The data were in hand, and their implication inescapable: we are at the center of a spherically symmetric distribution of gamma-ray-bursts sources.**[30] [emphasis added]

Is Geocentrism Proven?

Consider the implications of what's been discussed. Perhaps the theory that the universe began with a "Big Bang" 13.8 billion years ago is incorrect. And perhaps Earth occupies a central place in the cosmos and it does not revolve around the Sun. These would mark massive shifts in our thinking about the cosmos and where we live.

But in spite of the paradigm-shifting information discussed so far, there's still much to learn before it can be *concluded* that Earth occupies a special position in the cosmos. Robert Bennett, PhD, a geocentric physicist, summarizes the situation well: "How would you experimentally prove geocentrism, which means the Earth is at the center of the universe? Well, you would have to know what the boundaries of the universe are, in order to know that the Earth is equally distant in every direction from that boundary. Do we know the boundary of the universe? No. We know the boundary of the presently *observable* universe (or at least we think we do)....**So to say that the Earth is the geocentric center of the universe...is not really provable. What you can prove is that the Earth is on the axis of rotation of the universe.**"[31] [emphasis added]

Whether or not geocentrism is totally accurate is perhaps less important than uncovering what's *not* accurate. And based on the evidence discussed, there is good reason to question the validity of the mainstream, Big Bang–heliocentric model of the cosmos. Recall from the introduction that it's okay to be left in the position of saying "I don't know."

CHAPTER 3
EARTH'S MOTION AND THE AETHER

So far we've discussed geocentrism in terms of Earth's place in the cosmos, thereby challenging both the Big Bang model and the Copernican Principle that "we are not special." However, we haven't discussed, as explicitly, what the geocentric model implies about Earth's motion.

As mentioned at the end of the previous chapter, physicist Robert Bennett doesn't feel that geocentrism is "proven," per se. But he *does* believe that Earth is at rest. He calls this view *geostatism*.[1]

This implies that we are not moving: We are not moving around the Sun, and we are not rotating on an axis. Rather, we are stationary.

It implies that you, the reader, are not living on a moving object. As radical as it might sound due to our conditioning, it actually aligns with our everyday experience. We *feel* zero motion. We feel absolute stillness below us.

This plays out in areas of Earthly activity in which extreme precision is required. Surgeons who need true stillness to perform finely tuned surgeries don't worry about being jolted by a moving Earth. Tiny nudges could be a matter of life and death. Also, when we play the game Jenga, we don't worry that Earth will move and knock down the tower of blocks. In fact, in April 2023, there was a record-setting 101-inch stack of rocks that stayed balanced without falling.[2] No sign of motion there. Or consider other professions that require precision, such as watch- and clockmaking, microelectronics and semiconductor manufacturing, various forms of art and engineering, and astronomy.

Yet, on rare occasions we absolutely *do* feel the jolts from earthquakes. In everyday activities, however, we feel nothing.

According to the standard heliocentric model, we are moving in many ways—simultaneously—and really, really fast. We are told by scientists that even though we don't feel any of the motion, it's happening anyway. For example, you are allegedly moving right now because of the following phenomena:[3]

- Earth rotates on its tilted axis (23.5 degrees) at roughly 1,000 miles per hour (at the equator), which is faster than the speed of sound (767 miles per hour). The speed slows down going north and south from the equator until reaching the North and South Poles, where the rotational speed is nearly zero. So depending on one's latitude, the rotational speeds are very different.[4]
- Earth has multiple "wobbles" (known as the axial precession, nutation, libration, and polar motion, such as the Chandler Wobble and Annual Polar Motion).
- Earth revolves around the sun at 66,616 miles per hour. In fact, according to Albert Einstein's relativity theory, Earth is in a constant state of free fall around the Sun.
- Earth revolves around the Milky Way Galaxy at 514,000 miles per hour.
- Earth moves through space in an expanding universe at 1,342,000 miles per hours. Yes, you read that correctly: *you are allegedly moving more than one million miles per hour right now.*

We feel none of this motion. Absolutely zero. We experience pure stationary-ness. If scientists hadn't told us we're moving, we'd likely trust our senses and believe that we're living in a stationary place.

On an airplane or in a car or on a train or on a boat, we feel motion anytime there is even the slightest bump. We also feel changes in speed.

The rebuttal often arises: *We don't feel Earth's motion because it's constant.* But that's not true. For example, allegedly Earth has multiple wobbles, and its movement around the Sun is an elliptical path, which means it speeds up and slows down.

We're also told, in defense of heliocentrism, that Earth's motion is so gradual that we don't feel it—ever. But like most things in science, we're often given explanations as if they're indisputable facts. In some cases, scientists might be right. Not always, though.

Consider a thought exercise: Hot air balloons, helicopters, and drones float above the ground, sometimes for a really, really long time...and they come straight down. The ground doesn't move below them, leaving them behind. They don't have to adjust for how much Earth should have moved below. Apparently, as we're told, that's just because things in the air move with Earth's rotation. That's why airplanes allegedly don't need to account for Earth's rotation when they fly—they're just moving with Earth's rotation in the atmosphere above. And if they're flying on a trajectory that's not in perfect alignment with the direction of Earth's rotation, that's not a problem either, according to mainstream scientists.

As will become clear later in this book, there are many technical engineering documents stating that a stationary Earth can be assumed. **Even GPS (Global Positioning System) relies on "Earth-Centered, Earth-Fixed" technology.** And a manual describing the safe operation of tower cranes used for construction states: "Cranes must NOT perform slewing [that is, pivoting] and traveling motions simultaneously."[5]

Moreover, consider that in 2022, Shepard Humphries set the world record for the longest long-range rifle shot, hitting a target 4.4 miles away. *Outdoor Life* reported that Earth's rotation needed to be accounted for because it would cause the target to shift its position while the bullet was in flight (this is known as *the Coriolis Effect*).[6] However, cosmology researcher Jeran Campanella emailed Humphries, who told Campanella: "We did not take into account the Coriolis [Effect]," but he did mention other factors such as wind and temperature.[7] (Note: A knee-jerk reaction is often that the Coriolis Effect is proven because drains and toilets spin in opposite directions in the Northern and Southern Hemispheres. However, this is a misconception: the water can spin in either direction in both hemispheres because of factors such as the shape of the water basin and the angle of water entry, which impact the water's spin.[8])

The reality is that there is not any single experiment demonstrating that Earth moves: All existing experiments claiming that Earth does move could be explained by other phenomena having nothing to do with Earth's alleged motion. There are even studies that, according to some, debunk the belief in Earth's motion entirely. Albert Einstein, an avid heliocentrist, even acknowledged in 1922: "I have come to believe that the motion of the Earth cannot be detected by any optical experiment."[9] [emphasis added]

In the remainder of this chapter, I explore some of the key concepts and scientific evidence around the question of Earth's motion.

Foucault's Pendulum

Foucault's pendulum is often lauded as definitive proof that Earth rotates. *Smithsonian Magazine* summarizes the history concisely:

> On February 3, 1851, a 32-year-old Frenchman... definitively demonstrated that the Earth indeed rotated, surprising the Parisian scientific establishment. Acting on a hunch, Léon Foucault had determined that he could use a pendulum to illustrate the effect of the

> Earth's movement. He called together a group of scientists, enticing them with a note declaring, "You are invited to see the Earth turn." Foucault hung a pendulum from the ceiling of the Meridian Room of the Paris Observatory. As it swept through the air, it traced a pattern that effectively proved the world was spinning about an axis.[10]

The phenomenon allegedly demonstrates that Earth rotates, slowly, beneath the pendulum as it swings back and forth. This causes the pendulum's swinging direction to appear to change slightly over time. Its swinging pattern changes predictably and can be affected by latitude. Small objects are often placed in a circle around the pendulum and, as its swinging pattern changes, it knocks over the objects in accordance with Earth's alleged rotation.

However, the experiment's design raises questions because the pendulum requires some initial force (a "push") to make it swing in the first place. And that inherently biases the experiment, in spite of attempts to account for it. But more fundamentally, critics have claimed that the broader movement of the universe could be impacting the pendulum's motion. That is, from a geocentric perspective, the rotation of the stars around Earth could generate such an effect (more on this soon).[11] This perspective isn't often considered because mainstream science is so biased by heliocentrism. Heliocentrism is presupposed to be true, and from there, observations are made to fit that presupposition.

Even Einstein alluded to the possible impact of the broader universe on apparent local motion—explicitly naming Foucault's pendulum—in a 1913 letter to physicist Ernst Mach.[12] Furthermore, Brazilian physicist André Assis writes in his book *Relational Mechanics* (1999): **"Foucault's pendulum can no longer be utilized as proofs of the earth's real rotation."** His reasoning is as follows: "In relational mechanics, both facts can be equally explained with the frame of distant galaxies at rest...while the earth rotates relative to this frame, or with the earth at rest while the distant galaxies rotate around it....Both explanations are equally correct and yield the same effects. It then becomes a matter of convenience or of convention to choose the earth, the distant galaxies or any other body or frame of reference to be considered at rest."[13] [emphasis added]

Although the discussion above challenges the conventional explanation of Foucault's pendulum, it still presumes that Foucault's pendulum is indeed a robust effect. However, skeptics have pointed out that many modern Foucault's pendulums use electromagnetic drives to maintain the pendulum's motion.[14] This means that an external force is added to the picture, which introduces a potential distortion of the pendulum's natural movement. Drawing conclusions from such studies could be problematic. Additionally, some have questioned the replicability of the effect and have even claimed to demonstrate instances in which the pendulum did not behave as would be predicted by a rotating Earth.[15]

Stellar Parallax

The observation of *stellar parallax* is another line of evidence allegedly showing that Earth moves. *Parallax* refers to the movement of stars relative to background stars. Said another way, parallax can be seen by tracking one star in the sky and finding that it moves against the backdrop of other stars in the sky. From a heliocentric lens, these observations can be explained if Earth is in motion.

But if Earth is stationary, parallax would be explained by the movement of stars *around Earth*. This perspective is elucidated by physicist Robert Bennett. He notes in a 2023 interview with Austin Whitsitt: "All you can tell from parallax is relative position. You cannot tell whether something is objectively at rest or is moving, and that's an argument against what mainstream [science] claims."[16] He shows that the geocentric model, with a stationary Earth, predicts the same angle of parallax as the heliocentric model. Therefore, both models will agree on the amount of parallax—as long as you use the same "reference line" (which includes the Sun, the moving star, and the background star). As Bennett puts it: "You'll never see this in textbooks, which is the geocentric analysis of parallax."[17] The point here is: contrary to mainstream thinking, stellar parallax doesn't exclusively prove a moving Earth within a heliocentric model.

Additionally, heliocentrism must reconcile "negative parallax"—instances in which the stars move in the opposite direction from what would be predicted with a moving Earth. Negative parallax would appear to be a physical impossibility under heliocentrism, but the phenomenon is dismissed by mainstream science as the result of "measurement errors" or "statistical uncertainties."[18] Alternatively, this sort of explanation could be viewed as a mere excuse for anomalies that disprove heliocentrism, whereas they can be accommodated more easily if the stars move relative to a stationary Earth.

Stellar Aberration

In 1725, James Bradley found that the star Gamma Draconis was moving in an unexpected way. This movement is now known as *stellar aberration*. All stars exhibit such movement, and this is often regarded as proof of heliocentrism. Under this mode of thinking, the star appears to be in different locations in the sky *because Earth is moving.*[19]

However, as noted in 1901 by physicist Henri Poincaré, stellar aberration does not necessarily prove that Earth is moving within a heliocentric system: "The observation of the aberration shows us, therefore, not the movement of the earth, but the variation of this movement; **they cannot, therefore, give us information about the absolute motion of the earth**."[20] [emphasis added]

Airy's Failure and the Aether

Earth's motion has also been examined in conjunction with studies on the aether. More specifically, foundational studies in physics allegedly demonstrated that Earth does not move through the aether. But before looking into those studies, it's important to establish what the aether is.

The *aether* is an ancient concept, but it is now disregarded by mainstream physics. In the past, it was commonly believed to be a medium permeating all matter and space. It also served as a medium through which light traveled.[21] The aether was considered the "fifth element" (the other four were fire, air, earth and water) and was even important to the work of Nikola Tesla (1856–1943).[22] Some have speculated that the aether is a key to unlocking advanced propulsion technologies—or even "free" energy that could liberate much of the population (while hurting the energy industry's dominance).[23]

Additionally, there exists a 1950 Russian document, declassified by the CIA in 2011, titled "THE FAILURE OF US ATTEMPTS TO ATTAIN SUPREMACY OF THE ETHER."[24] So one might wonder if, at some level of authority, the aether has persisted as a secretive scientific reality.

However, interpretations of pivotal 19th-century studies steered mainstream physicists *away* from the reality of the aether. That leaves us where we are today, in which it is not part of typical thinking. More specifically, the bias toward a *moving Earth within a heliocentric system* has steered modern physics away from the aether.

A common theme has emerged, and it will be discussed in the coming examples: Physicists decided that certain experiments disproved the aether and preserved the notion that Earth moves. However, they could have interpreted the results as being validation of the aether within a stationary Earth. To simplify, the options were:

- **a. No aether and a moving Earth**
- **b. The aether and a stationary Earth**

Physicists chose the former rather than the latter. And it stuck.

For instance, physicist and mathematician George Airy conducted a study in 1871 that is widely regarded as a failure to detect that Earth is moving through the aether. In the experiment, Airy viewed a star through two telescopes: one filled with air, and another filled with water. He thought that the starlight would not hit the water-filled telescope's eyepiece directly, whereas this issue wouldn't occur in the air-filled telescope. He thought this because (a) he knew that a property of water is to slow down light, and (b) because he assumed Earth was moving. Given these assumptions, the water in the telescope would slow down the starlight while the telescope would be moving along with Earth—and the resultant delay would change the angle in which the starlight would hit the telescope's eyepiece. To compensate, Airy thought he needed to tilt the water-filled telescope at a different angle relative to the air-filled telescope. **He found, however, that the tilt of the telescope didn't make a difference.** When the air-filled telescope and the water-filled telescope were pointed at the same angle, the starlight hit the telescope's eyepiece *in the same way*.

Yet instead of concluding that "Earth is stationary," the scientific community has interpreted the results to mean that "Earth is not moving through the aether." **Conversely, the study's results can be explained with a stationary Earth.**

But what about the aether within this explanation? As physicist Robert Bennett remarks, an "aether wind" would slightly alter the starlight from above—for both telescopes. This would result in a necessary tilt for both telescopes to account for the aether wind (and also for the star's slight movement in the sky, known as *stellar aberration*). In Airy's study, both the air-filled and the water-filled telescopes were tilted the same way and got the same result. This was suggestive of an aether wind that *equally affected both telescopes* within a stationary Earth. (Note: See Austin Whitsitt's 2023 interview with Robert Bennett, PhD [Part I] for a helpful animation that makes the experiment easier to understand.[25])

In Bennett's view, the results had convincing implications:

> [Airy] proved two critical things: Earth is stationary and not only is there an aether, but it moves. And in fact it affects starlight by pushing it sideways....This should have put an end to the discussion. But the tendency of mainstream [science] is to ignore any experiments that disprove what they think. So they continued to think 1) that Earth is moving and 2) either that there is no

> aether, or that if there is an aether it's rigid like metal and it's stationary and there is no such thing as an aether wind or an aether drift....To this day, in 2023, from 1871, we still have no proof offered by scientists that will disprove what is called "Airy's Failure." **It's actually "Airy's Success." He successfully showed that Earth is at rest and that there's an aether wind surrounding the Earth that's pushing all of the starlight....[This is] a very important experiment and it's usually totally ignored. If you [heard] this as part of a physics class, you were very lucky because I never saw it shown.**[26] [emphasis added]

Physicist Hendrik Lorentz put it well in 1886 when he commented on Airy's experiment: **"Briefly, everything occurs as if the Earth were at rest."**[27]

Michelson (1881) and Michelson-Morley (1887)

In 1881, physicist Albert Michelson ran another study designed to detect Earth's movement through the aether. He shot light beams in two directions: one west (the direction of Earth's alleged motion around the Sun), and another toward the North Pole. The westbound light beam would have been expected to bump up against the aether in its movement with Earth around the Sun, whereas the northbound light beam shouldn't have had as much of an issue. Robert Bennett compares it to swimming directly against the current versus swimming across (perpendicular to) the current.[28] In other words, the westbound light (swimming directly against the current) would be expected to move more slowly, as measured by an interferometer. (Note: This is a simplified description of the experiment.)

But Michelson found that the difference between the two light paths was *minimal rather than significant*. Scientists would have expected a significant difference if Earth's motion had impacted the westbound light's motion. **Thus, Michelson stated in his paper published in the *American Journal of Science*: "This**

conclusion directly contradicts the explanation...which presupposes that the Earth moves."[29] [emphasis added]

Michelson ran the study again in 1887, this time with Edward Morley. They used a more sensitive interferometer and tried to eliminate disturbances that could pollute the experiment. **They got the same basic result as Michelson did in 1881.[30]**

Instead of proclaiming that the studies disproved Earth's motion, the studies are known for having "disproved the aether" because of a "null result." Scientist John Bernal even wrote in 1969: "The Michelson-Morley experiment was the greatest negative result in the history of science."[31]

But, as Bennett notes, they didn't quite get a "null result."[32] Rather, there was a detection of the aether beyond experimental error because the light beams weren't perfectly aligned (the experimenters found a "fringe shift"). **Thus, contrary to the mainstream conclusion that Earth is moving and there is no aether, the study could be interpreted to suggest that Earth is *stationary* and that an *aether exists*. This alternative explanation of the Michelson-Morley study is often ignored by modern physicists, ultimately pointing us to our current cosmological model in which 96 percent of the universe is a mystery and there's no unifying theory of physics.**

And, as noted by Bennett, physicist Dayton Miller ran experiments with a similar design and further validated the notion of "an aether wind that's coming from stars."[33] Yet the notion that the aether *was* detected is quite different from what we hear from the mainstream, such as *Britannica's* summary claiming that Michelson-Morley's "null result seriously discredited the ether theories."[34]

The idea that Earth's motion, or lack thereof, was studied in the experiment is also often ignored by mainstream physicists. The focus is typically on "disproving the aether." Yet the Michelson-Morley paper itself was titled "On the Relative Motion of the Earth and the Luminiferous Ether"—meaning that *Earth's motion* was clearly a critical part of the study. [underscore added]

It's also important to emphasize how many prominent thinkers *have* acknowledged the true implications regarding Earth's motion. What follows is a series of quotations to this effect, compiled by Robert Sungenis:

- Physicist Henri Poincaré wrote in his 1901 book *Science and Hypotheses*: "A great deal of research has been carried out concerning the influence of the Earth's movement. The results were always negative."[35]

- Poincaré remarked again, this time in 1905 in *The Monist*: "Are we about to enter now upon the eve of a second crisis? These principles on which we have built all, are they about to crumble away in their turn?… Alas…such are the indubitable results of the experiments of Michelson."[36]

- Physicist G. J. Whitrow stated in 1949: "It is both amusing and instructive to speculate on what might have happened if such an experiment could have been performed in the sixteenth or seventeenth centuries when men were debating the rival merits of the Copernican and Ptolemaic systems. The result would surely have been interpreted as conclusive evidence for the immobility of the Earth, and therefore as a triumphant vindication of the Ptolemaic [geocentric] system and irrefutable falsification of the Copernican [heliocentric] hypothesis."[37]

- Nobel Prize–winning physicist Wolfgang Pauli commented in 1958 on the Michelson-Morley study and dozens of others performed after it, confessing to what he called "the failure of the many attempts to measure terrestrially any effects of the earth's motion."[38]

- Physicist Bernard Jaffe wrote in his 1960 book, *Michelson and the Speed of Light*: "The data were almost unbelievable.…There was only one other possible conclusion to draw—that the Earth was at rest." Jaffe demonstrated his Copernican bias when he then remarked, "This, of course, was preposterous."[39]

- Physicist James Coleman wrote in his 1954 book *Relativity for the Layman*: "The easiest explanation was that the earth was fixed in the ether and that everything else in the universe moved with respect to the earth and the ether....Such an idea was not considered seriously, since it would mean in effect that our earth occupied the omnipotent position in the universe, with all the other heavenly bodies paying homage by moving around it."[40]
- Physicist Adolf Baker wrote in his 1960 book, *Modern Physics and Antiphysics*, that Michelson-Morley's results "suggested that the Earth must be 'at rest.'"[41]
- Physicist Richard Wolfson remarked in a series titled *The Great Courses*, released in 2000: "What's the conclusion from the Michelson-Morley experiment? The implication is that the earth is not moving."[42]

The point is: Many prominent physicists have made profound statements about the studies' results. But it's worth noting that in spite of these results, Michelson himself was a heliocentrist and thus believed in a moving Earth. In fact, he conducted another important study in 1925, known as the Michelson-Gale-Pearson experiment. It examined Earth's alleged rotation on its axis (using the "Sagnac Effect"), rather than examining Earth's motion around the Sun. As Bennett comments, the same sort of interpretive error occurred when the results of the study came in: "rotation of the aether was interpreted as the rotation of the Earth."[43] Yet again, the results could have been applied as demonstrations of a stationary Earth, but they weren't. (Note: Along these lines, Bennett likewise observes that Foucault's pendulum could be explained by the aether rather than Earth's rotation.[44])

Einstein and Relativity Theory

Michelson's work even impacted Albert Einstein. As Einstein stated in a 1922 speech in Kyoto: "While I was thinking of this problem [that is, the motion of Earth] in my student years, I came to know the strange result of Michelson's experiment. Soon I came to the conclusion that our idea about the motion of the Earth with respect to the ether is incorrect, if we admit Michelson's null result as a fact. This was the first path which led me to the special theory of relativity."[45] Einstein's biographer Ronald Clark also acknowledged that "by the time that Einstein came on the scene other experiments had been carried out," and they had similar results to Michelson's.[46] [emphasis added]

Einstein continued in his 1922 Kyoto speech as follows: "Since then I have come to believe that the motion of the Earth cannot be detected by any optical experiment, though the Earth is revolving around the sun."[47] Here, he made a startling admission about the apparent inability to detect Earth's motion, but he retained his heliocentric view. The Copernican Principle held strong. [emphasis added]

It has remained so strong, in fact, that Einstein's world-changing theory of relativity was perhaps created in order to preserve it. Some might even call relativity theory a post hoc rationalization. Basically, Einstein devised a framework in which Michelson's results could be explained with a moving Earth. **He contended that the measuring device in the studies *shrank*, which was enough to mask Earth's movement around the Sun.** (Note: Einstein's notion of length "contraction" bore similarities to the prior work of physicist Hendrik Lorentz.[48])

A cynical perspective of Einstein's work is as follows: *Einstein had to say the measuring device shrank in order to preserve the mainstream cosmological model. So he invented an entirely new theory. And now, we're stuck with a fallacious view of the cosmos that has to be propped up by dark matter and dark energy, which might not even exist.*[49]

Under Einstein's cosmology-altering idea, not only do objects shrink when they move, but time slows down. Einstein's initial theory is called the "special theory of relativity" (1905) and was later broadened to explain gravity as "the curvature of space-time" with the "general theory of relativity" (1915).[50] **These concepts**

are now foundational to modern physics, and they are linked to Einstein's assumption that Earth moves.

But if Einstein misinterpreted Michelson's results, what would that mean for relativity theory? Sungenis summarizes the situation:

> Einstein realized that if Michelson's experiment detected any ether, even a tiny bit, then the theory of Special Relativity—which was Einstein's only answer to Michelson's experiment—would be immediately falsified. Ether, even a tiny bit, would provide an absolute frame from which to measure velocity, and the theory of Special Relativity had divorced itself from all such absolute frames.
>
> This fact was admitted in a conversation Einstein had with Sir Herbert Samuel in Jerusalem. Einstein stated: "If Michelson-Morley is wrong, then Relativity is wrong."
>
> In other words, if Michelson's experiment was not null and ether existed, then Einstein's theory of Special Relativity was categorically wrong. **As we can see, everything hinged on the proper interpretation of Michelson's experiment.** Consequently, the critics were vociferous in pointing out that Einstein ignored the fact that Michelson's experiments always yielded a small positive result, and thus his theory of Special Relativity was, indeed, wrong.[51] [emphasis added]

Stated another way by Columbia University physics professor Charles Lane Poor in 1922: **"The Michelson-Morley experiment forms the basis of the relativity theory: Einstein calls it decisive. If it should develop that there is a measurable ether-drift, then the entire fabric of the relativity theory would collapse like a house of cards. For this reason the repetitions of the Michelson-Morley experiment recently made at Cleveland and [Dayton Miller's studies] at Mount Wilson are of special importance: they indicate that...there may be a measurable ether-drift."**[52] [emphasis added]

If relativity theory were to collapse like a house of cards, as Poor suggested, then modern cosmology would truly need to start from scratch.

But the problems for Einstein's relativity theory don't end there. In particular, his treatment of light has brought about further criticism. A core tenet of special relativity is that the speed of light is constant (in a vacuum),[53] but this seems to create an inconsistency with his broader general relativity in which the speed of light is not constant. Einstein mentioned this in his book *Relativity: The Special and General Theory* (written in 1916): "Our result shows that, according to the general theory of relativity, the law of the constancy of the velocity of light in vacuo [that is, in a vacuum], which constitutes one of the two fundamental assumptions in the special theory of relativity...**cannot claim any unlimited validity**."[54] [emphasis added]

And yet special relativity is a subset—a special case of—general relativity. **As Robert Bennett summarizes the apparent incompatibility of Einstein's special and general relativity: "The two [theories] should match, but they can't."**[55] His critique of Einstein and relativity is particularly noteworthy because his PhD thesis was written on the topic of general relativity, and he was surrounded by many "relativity nuts," in his words—including his very own thesis adviser whose mentor worked directly with Einstein.[56]

Rethinking Gravity

If Einstein's relativity theory is indeed as fundamentally flawed as just discussed, then there's some deeper reconsidering to do. That means questioning the fundamental concept of gravity.

Gravity was conceived by Isaac Newton in the 1700s, and Einstein's theory of general relativity is often regarded as a more accurate version. Although Newtonian ideas are colloquially regarded as "law," what's less often discussed is where the theory fails. **For instance, Newtonian gravity cannot accurately predict Mercury's motion.**[57] A theory that fails to predict something so fundamental, no matter how predictive it is in other local circumstances, is, by definition, defective. And we already discussed the problems with general relativity, which is supposed to be a more accurate version of gravity.

A "house of cards" comes into clear view. Gravity and general relativity have led physicists to conclude that 96 percent of the universe is made of unexplained dark matter and dark energy. As discussed in the introduction, these entities might not even exist. They might simply be *assumed* to exist to maintain a Big Bang–heliocentric model that's predicated on gravity, general relativity, and a moving Earth. But recall that astrophysicist Pavel Kroupa and his colleagues have established with high confidence that dark matter has now been falsified, and the 2006 "Report of the Dark Energy Task Force" likewise described serious problems with dark energy.

From this perspective, it's no wonder that we don't have a unifying theory of everything in physics that can reconcile the two leading theories (relativity and quantum mechanics). As summarized in a 2023 article in *iai News*, written by Kroupa and his colleague Moritz Haslbauer: "Our model of the universe has been falsified: The cosmological standard model is wrong.... Dark matter and dark energy are speculative, and a range of recent observations...are increasingly showing to anyone willing to see that the universe doesn't look or behave the way the standard cosmological model predicts. It's about time the cosmology community gave up on this theory rather than digging itself into a deeper hole filled with speculation and fantasy."[58] [emphasis added]

With this in mind, let's look at the theory of gravity more closely. Gravity is basically the idea that "mass attracts mass"—that there is a force within mass that brings other mass to it. An apple falls to the ground because the mass of Earth is greater than the mass of the apple, so the apple moves to the ground.

However, can we know for sure that "mass" is where the force is coming from? Does the mass of Earth cause the apple to fall?

Surely objects fall to the ground. But that doesn't mean a force called gravity is necessarily the reason for it. There could be other explanations.

We have to wonder: *Has gravity been proven as the true cause of our observations?*

Steven A. Young, who holds a master of science and PhD in theoretical physics, similarly critiques gravity in his 2024 book, *A Fool's Wisdom*:

> The original measure of gravity, which is alleged to have proved its existence, is called the Cavendish experiment (1797), a famously difficult experiment to conduct due to the extremely sensitive nature of the apparatus. [M]easurements are so tiny that almost anything could be causing the readings. It's a similar fallacy to the so-called LIGO [Laser Interferometer Gravitational-Wave Observatory] gravitational wave detector; the sensitivity of the instrument is so incredibly precise that any minuscule blip on a graph can be blown out of all proportion and interpreted as evidence of a "galactic gravity wave" or "supernova collapse" or some such thing. These alleged "proofs" are absolutely feeble.[59]

Young continues: "The observed effect of the falling apple does not imply the existence of 'a force called gravity' as the cause. There is no need to explain the phenomena of falling by invoking a fictitious force. **Density and buoyancy are sufficient principles to explain the rising and falling of objects, as well as electrostatic attraction and repulsion."**[60] [emphasis added]

Density and buoyancy are easy to explain: denser things fall and lighter things float; a marble falls to the floor when dropped because it's denser than the air around it, and a helium balloon rises because it's less dense than the air around it.

But why do denser objects fall down rather than move in some other direction? This "downward bias" of denser objects relates to Young's mention of an "electrostatic" force.

Caltech physicist Richard Feynman discussed this in a lecture series in the early 1960s, stating that "a small current—caused by the electric field...—passes from the sky down to the earth."[61] NASA even has an "Electrostatic Levitation Laboratory."[62] Similarly, the Hutchinson Effect, uncovered in 1979 by John Hutchinson, includes phenomena like the levitation of heavy objects and involves electrostatics.[63] In a sworn affidavit he submitted for a legal case in the Southern District of New York, Hutchinson mentions that his work "has been classified as a matter of National Security by the Canadian Security Intelligence Service." He includes in the affidavit a 1986 letter from the Canadian government to this effect.[64] (Note: For further information on these technical matters, see Austin Whitsitt's YouTube presentation made on January 10, 2024. He references many studies in which thunderstorms affect the rate at which objects fall, which in his view lends support to the electrostatic concept. The question is not whether things have been observed to fall at a similar rate, approximately 9.8 m/s^2, but rather what *causes* that to happen.[65])

But even from a less technical perspective, there are reasons to question the traditional narrative around gravity. For example, in "outer space," we are told that there are spheres revolving around spheres—moons revolving around planets and planets revolving around the Sun. And we are told that this occurs because of gravity (more specifically via Einstein's general relativity theory). But while here on Earth, have we ever shown that the mass of one sphere causes another sphere to revolve around it? Certainly, we're told that the phenomenon of spheres revolving around spheres occurs on big scales because gravity is such a weak force and requires huge masses in order to see certain effects. It's noteworthy that a phenomenon we see "out there" isn't what's observed "in here" on Earth. So, scientists merely assume that gravitational effects explain what they see out there.[66]

Consider another example: Earth has water all over it. The water is said to stay there, rather than fall off, because of gravity. But can we ever make water stick to a sphere in an experiment on Earth? Try pouring water on a soccer ball and see if it sticks in the same way that oceans are said to stay on Earth.[67]

Critics of the conventional theory of gravity question the narrative that massive bodies of water, such as oceans, can stick to Earth. In everyday living, this phenomenon is not observed: water falls off of spherical objects.

And if a wet tennis ball spins rapidly, the water flies off of it. When Earth spins (allegedly), water does not fly off.[68] We're told that this is because Earth has such a large mass that gravity keeps the water here. But is that true?

Even the notion that gravity causes tides on Earth—via the *equilibrium theory* that balances the gravitational effects of the Sun and Moon—is flawed. The Virginia Institute of Marine Science calls it "an example of a model of ideal behavior—something that works for the purpose intended although it may not adhere to the truth in all instances....Observations of real tides show that they do not respond instantly to the tide-producing forces of the moon and sun as the theory requires."[69] Maybe other forces are at play that need to be considered.

In summary, the allegiance to gravity has huge implications for all of cosmology. As Young remarks: "To combat the blatant failures of the gravitational theory, physicists have postulated new forces that are alleged to compensate for gravity's destructive all-sucking effect, things such as 'anti-gravity' or 'dark matter,' which are said to work against the attractive force to balance everything out again so it all feels perfectly calm and stationary like we observe. This is fiction on top of fiction, they are attempting to compensate for the absurdity of the original idea by inventing even more absurd ideas."[70]

But if everything *is* "perfectly calm and stationary"—because Earth is not moving—then some serious rethinking needs to happen about *everything*. Thus, in the spirit of "starting from scratch," it's appropriate to revisit not only Earth's place in the cosmos, but also what it fundamentally is. That exercise begins in the next section.

PART II
WHAT IS THE NATURE OF OUR HOME?

CHAPTER 4

SKEPTICISM ABOUT THE NATURE OF OUR HOME

So far, my analysis has focused on questioning Earth's place in the cosmos and its alleged motion, whereas my attention now will turn to *questioning the nature of Earth itself.*

If we've misunderstood the nature of our home, one might wonder: *Are there land masses that we do not know about? Are there valuable, hidden resources that could reduce scarcity; make life easier; and contribute to our freedom? Are there aspects of human history that are being misunderstood? Are there aspects of our very identity that are being misunderstood? Might there be people in positions of power who want this information to remain hidden?* Ultimately, getting to the truth of where we live is an important prerequisite to decoding what we're doing here.

As a warning, the forthcoming discussion truly requires letting go of all past belief systems. Hopefully now that the basic cosmological model has been put into question, the ideas will be slightly more tolerable. However, they can be quite challenging and destabilizing—especially if this is your first time encountering these concepts.

As you read, try to keep in mind that the Big Bang, heliocentric cosmology has shaky foundations. Some would argue that it has been downright falsified already. There is good reason to question whether Earth moves, and basic concepts such as gravity and relativity theory likewise have reason to be questioned. Not only that, but we're told that 96 percent of the universe is unexplained dark matter and dark energy and there exists no unifying theory of everything.

Doesn't it then seem reasonable to rethink the fundamental nature of where we live? It is from this foundation that I will discuss the nature of Earth over the next three chapters.

The current chapter (4) explores an alternative framework for where we live—one that has a long history and has seen momentum recently. Unfortunately, however, it is currently impossible to develop a comprehensive model or map of where we live because of a limited ability to know what's above and below us, and because of travel restrictions (most notably in Antarctica).

In chapter 5, I focus on what's *not true* about where we live rather than trying to prove what *is* true. In other words, I explore pieces of evidence that directly challenge the current model of Earth. Recall from the introduction that no definitive replacement model needs to be proposed: falsifying one model does not necessitate coming up with a conclusive alternative. That leaves us in a place of saying "I don't know" about the nature of where we live. But at least the exercise can narrow down the set of possibilities.

In chapter 6, I explore the remaining elephant in the room: "What about space agencies?" Sadly, a close examination of NASA's history—including its Nazi and occult origins—gives real reason for skepticism. This includes questioning the Apollo Moon missions. NASA admits that almost all of the missions' telemetry tapes no longer exist. More recently, NASA has admitted to doctoring its images by using photoshopping, artists' renderings, fish-eye lenses that show artificial curvature, computer-generated images, and composite images. For instance, NASA's Earth Observatory website states that the "spectacular 'blue marble' image [of Earth]" is based on observations that were "**stitched together**."[1] Consider this: there exists no continuous, unedited video footage from outer space of the entire Earth, showing its full rotation and all of its geographic locations. And, likewise, there exists no single photograph showing Earth in its place within the alleged solar system. We're just shown snippets and asked to extrapolate, or we're shown images that have been doctored. This is just the tip of the iceberg in terms of what is covered in chapter 6. [emphasis added]

Questioning the Globe Model

Let's start with the basics. According to mainstream cosmology, Earth is spherically shaped. More specifically, it is an oblate spheroid that bulges at the equator due to the "force of gravity," combined with its alleged rotation. Neil deGrasse Tyson colloquially referred to Earth as pear-shaped.[2] In other words, as it's often put, Earth is a "spinning ball flying through space." That's where we live. From now on, I will refer to this model as "the Globe."

But some skeptics question Earth's shape, alleging that it is a topographical plane with peaks and valleys. More casually put, it is "flat." In other words, all of the land masses and water in the spherical model are simply part of a plane rather than a globe. Also, under this line of thinking, Earth is stationary and occupies a central place in the cosmos. This general concept—known as "Flat Earth"—harkens back to an ancient cosmology held by cultures all over the world.[3]

There's often a knee-jerk reaction against the notion of a Flat Earth because of the modern stigma. However, it's often the case that critics don't even understand what a Flat Earth entails. That's why, in this chapter, I will delve into the framework itself—to at least bring us to a shared idea of what the alternative theory is—regardless of whether the ideas are found to be compelling.

For instance, detractors might instinctively claim that a Flat Earth can't possibly exist because it would collapse on itself due to the force of gravity. Therefore, a sphere is the only physical possibility for the shape of our home. However, that argument presupposes that gravity exists, whereas we've discussed that perhaps it does not exist (at least not in the way it's commonly understood).

Although one might think that it would be easy to debunk other aspects of a Flat Earth, the matter isn't so straightforward. This raises an important question to keep in mind: if you wanted to prove that the Globe is the correct model, how would you go about doing so? Along these lines, consider the following statement made by Robert Sungenis, a geocentrist who believes that Earth is spherical rather than flat. In his 2018 book, *Flat Earth/ Flat Wrong*, he writes: "The task of proving the Earth is not flat, at least from ground level, is not always as easy as one might think, and in some instances the evidence is equivocal. In reality, just as cartographers can make many kinds of maps of the Earth, so one can make a flat Earth model fit into many of the things we see on Earth. This doesn't mean, however, the Earth is flat. It only means that sometimes the evidence is pliable and/or missing certain crucial ingredients."[4]

Another common misconception is that anyone who questions the Globe automatically believes that Earth is flat. That's not true. Many people hold an opinion along these lines: *I don't know what shape Earth is, but we're definitely not living on a spinning ball flying through space.* This position is intellectually honest because there are too many unknowns to create a comprehensive and perfectly precise model of our home. For the remainder of this book, I will use the general term *Globe Skeptics* to refer to those who question the Globe. The point is that even if the Globe is not accurate, Earth might simply be "something else" that hasn't yet been uncovered. Perhaps the binary notion of "Globe" or "Flat" creates a false dichotomy and leaves out other creative possibilities that either aren't being considered or are beyond current human comprehension.

Along these lines, consider the following statements from Steven A. Young—who, as a reminder, holds a master of science and a PhD in theoretical physics. He publicly came out as a Globe Skeptic in his 2024 book, *A Fool's Wisdom*:

> I first heard of "flat earth" around 1999....Of course, my student friends and I all laughed at the idea of people who believe the earth is flat, how could they be so stuck in the past?...I didn't think about it again for many years. During 2015–2016 it popped up a number of times online, and eventually by recommendation from a friend, so I decided to see what it actually is these people believe. I expected it to be some weird cultish thing like Scientology, but what I found was just solid, verifiable scientific research that calls into question many of the theories we are taught about our world. For the most part there was no nonsense at all, just straight-shooting facts and hard questions that are thought provoking and irrefutable. As someone coming from a scientific background, I was well impressed by how meticulous and well-reasoned the flat earth research was; it was a breath of fresh air to see the scientific method being used so effectively. I couldn't understand why I hadn't learned any of this in university; it's really fundamental stuff. It took a few months of watching presentations and doing experiments and having many heated conversations with people, **but eventually I came to a place of 100% knowing that our world is not a planet spinning through space. I may not know 100% what it is, but I know 100% what it is not, and it's not a spinning ball.** [emphasis added]

Young elaborates on his journey as well as the challenges of getting academics to engage:

> The process of conversion is a one-way street; nobody ever goes back to believing in the ball, there are no "ex-flatearthers," as far as I know, because there is actually no reason to believe you are spinning once you know that you are not. But how does one get to a place of knowing such a thing? Well, start by asking yourself, how do you know that you're on a spinning ball flying through space? **You have to set about trying to**

prove the globe, something that seems like it should be quite an easy task...but it turns out [it] is actually quite impossible. Many will tell you the same story: becoming a "flat earther" is the result of having tried to prove the globe and failed. The irony is flat earthers are treated as the most stupid people in society, a proper fool..., a symbol of ignorance, madness, and wrong-think, a group of people not even worth listening to. The globe is so entrenched in the collective consciousness that it is generally deemed to be an unquestionable fact of reality. To do the work it requires a position of neutrality, no attachment to any particular outcome, just simply to look at all the evidence for and against the spinning ball earth....I remember vividly the day I realised I wasn't spinning through space; it was the most profound awakening of my life, a true enlightenment experience, a new dawn, finally getting "out of the mind" and "coming to the senses"....[emphasis added]

When I set about trying to prove [the globe] back in 2016, I thought it would be child's play, with my trusty PhD and advanced mathematical skills, and since I was flying "around the globe" in planes all the time, I could undoubtedly prove it, easily! But the more I looked and the harder I tried, I just couldn't. I searched for globe evidence every single day for many years, it was an obsession, I kept thinking maybe there was something I overlooked, some test that proves it....I followed the debate online closely, but I have never found a single reason to believe in the globe in all the years since. The case for the globe gets weaker with each passing year, while "flat earthers" grow in number constantly.

What's clear is that academics don't engage with it on a sincere level; they either completely ignore it, or use *ad hominem* attacks to slander and belittle the researchers, as if the whole topic is beneath them, an insult to their intellect, a conversation not worth

> **having. This is intellectual bypassing, they just block you and smear you and carry on as normal.**[5] [emphasis added]

Young is one of many professionals to come forward with such beliefs, including pilots and various scientists. David Weiss's Flat Earth Sun, Moon & Zodiac Clock app, at https://theflatearthpodcast.com/, includes videos of such professionals, among many other basics related to Flat Earth concepts (in particular, see the Frequently Asked Questions playlist of videos under the heading "Pilots and Scientists?").

Young does not mention the exact nature of the tests he ran during his "conversion" process, but many Globe Skeptics use a camera, a telescope, or binoculars to discover whether they can observe distant objects that should be blocked by the physical curvature of the Globe. It would be like trying to see something that's around a bend—it's impossible because there's a physical structure in the way. This is a simple way to test the Globe. But all too often, people claim that they can see such objects—and this is likely one of the reasons why Globe Skepticism is on the rise. In fact, instances like these were presented in a United States court in a case won by a Globe Skeptic in 2019.

In the next chapter, I will discuss many additional phenomena that Globe Skeptics feel challenge (or even outright debunk) the Globe, such as long-distance laser tests; long-distance mirror flash tests; ships disappearing over the horizon due to optical limits and then coming back into view when using a zoom-lens camera; the horizon rising to eye level with altitude rather than curving down and away from the viewer; horizontal wave propagation over distances that shouldn't be possible on a sphere; whale sonar communication over distances that shouldn't be possible on a sphere; specular (mirrorlike) reflections over long distances that are indicative of a perfectly flat water surface; anomalies with shooting stars, meteors, and asteroids; navigation problems arising from magnetic declination (resulting in corrections needed for the difference between the Globe's geographic and magnetic poles); the existence of nineteen time zones in the Northern Hemisphere versus thirty-two time zones in the Southern Hemisphere and twenty-four at the equator; and selenelion (lunar) eclipses in which both the Sun and the eclipsed Moon are visible at the same time (which should not occur if Earth is causing the eclipsed Moon by sitting in between the Sun and Moon).

The point is that much more thought has gone into Globe Skepticism than many people assume.

The General Flat Earth Framework

Although a framework is often put forth for a Flat Earth, many aspects of it aren't even uniformly believed by Globe Skeptics. And there certainly isn't much funding to evaluate such an alternative view. On top of that, as I will discuss later in this chapter, there is likely an infiltration of intentional disinformation that misrepresents and discredits the position. It's a "poisoning of the well," of sorts. Globe Skeptics are often censored on the internet too. Thus, they are working from a rough position, especially when having to re-explain basic belief systems about all of physics and the cosmos.

There can also be a tendency among Globe believers to find one thing they can't conceive of on a Flat Earth and then shut down their entire exploration. However, this reflects a double standard, because these individuals are often willing to excuse massive problems with their own Globe model—such as the unexplained 96 percent of the universe—and just sweep them under the rug as excusable anomalies. One black swan destroys the model claiming that "all swans are white"—and that's what is important in this exercise.

With that context in mind, the basic Flat Earth framework goes something like this:[6]

Earth is *not* a disc floating through space revolving around the Sun. Rather, what we know of as Earth occupies a central position in the cosmos and looks like a massive circular lake surrounded by an ice-wall perimeter. This implies that Earth is round in a *circular* sense, which is distinct from being round as in a *spherical* Globe.

All of the land masses and bodies of water within Earth are contained in this lake, and the ice-wall perimeter is what the Globe model calls "Antarctica." In other words, on a Flat Earth, Antarctica is not an island continent. It is the circular enclosure of Earth. Thus, the Flat Earth does not have a nearby "edge" that people would fall off if they went too far. They'd simply hit the ice wall. But what's beyond the ice wall is left for speculation. Many wonder if there are valuable resources and other massive worlds extending well beyond the ice wall, but no one knows with certainty because public exploration of Antarctica is so limited.

It's also worth noting that even by the Globe model's standards, Antarctica has the highest average elevation of any continent on Earth. Antarctica was described in 1894 by General A. W. Greely, an American explorer, as an "ice barrier [that] is the fore-front of the enormous glacier-covering."[7]

Many different Flat Earth maps and illustrations have emerged throughout history, and there are differences between them. So there isn't a single map that is known to be "the Flat Earth map." Sometimes the 1892 Gleason map is used to represent the Flat Earth as an approximation, but some Globe Skeptics (such as world-renowned sailor Hervé Riboni) view it to be imprecise.[8] Yes, you read that correctly: a world-renowned sailor is a Globe Skeptic. The Gleason map is shown here for reference.

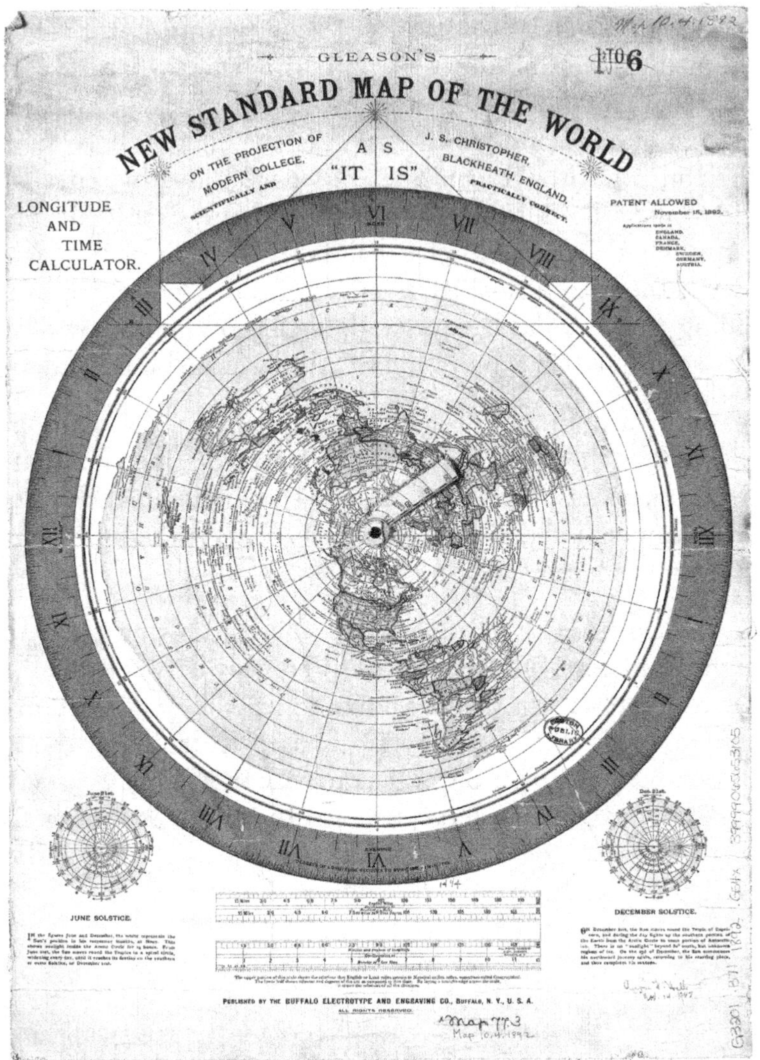

The 1892 Gleason Flat Earth map is a commonly presented version of a Flat Earth map.

Even using that imperfect model, some Globe Skeptics claim that certain flight patterns seem strange on a Globe but make more sense on a Flat Earth. Eddie Alencar's book *16 Emergency Landings Proving Flat Earth* (2019) uses flight data to compare emergency landings on a Globe versus the (imperfect) Flat Earth Gleason map, and suggests that the Flat Earth can explain the examples better.

Moreover, Globe Skeptics often note that logos of various international organizations show the same basic Flat Earth shape. Some feel this is evidence that the Flat Earth is known to be true by certain members of elite power structures. Others say that the logos merely reflect innocent two-dimensional illustrations of a spherical map, so we shouldn't take them literally. They are shown below for reference.

Many organizations' logos are two-dimensional images of Earth that some construe to be renditions of a "flat" Earth.

Circumnavigation

The center of the circular lake of Earth is the magnetic North Pole.[9] That means moving directly away from the North Pole in any direction is moving "South," in which case one would eventually hit the ice-wall perimeter (Antarctica). Consequently, the Flat Earth does not have a "South Pole." There's only a North Pole.

Moving due East or West relative to the center point of the North Pole takes one on a *circular* path. In other words, a traveler would need to maintain the same distance from the North Pole at the center, and that entails going in a circle until returning to the starting point. Thus, circumnavigation is easily explained on a Flat Earth.

Yet, consider physicist Carlo Rovelli's comment in his 2011 book, *Anaximander and the Birth of Science*: "As much as one may wish to believe that the Earth is flat, the day comes when he must face the fact that Ferdinand Magellan's ship set off toward the West and came back from the East [in the 16th century]."[10] This statement reflects a lack of understanding of the basic Flat Earth concept, which is typical of the way many mainstream critics approach the issue. All Magellan's ship would have had to do was set the compass to "West," which would have led the ship on a circular path around the North Pole at the center, eventually returning to the original point. But this doesn't prove a Flat Earth. Rather, it demonstrates that both a Flat Earth and a Globe can explain basic circumnavigation. It doesn't exclude or prove either one.

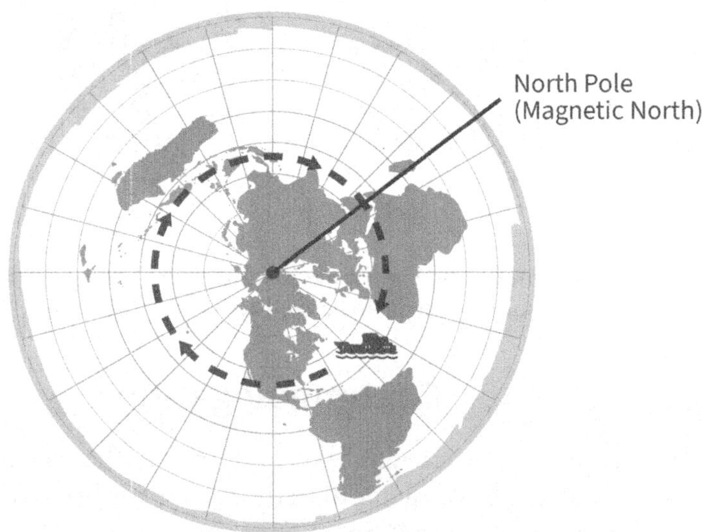

In a basic Flat-Earth map, going due East or West entails going in a circle because North is at the center point. So, continuing to go East or West requires constant turning relative to the North Pole such that the navigator will end up back where he or she started, eventually. Circumnavigation is thus explained by a Flat Earth (and, of course, it's also explained on a Globe).

Day/Night Cycles and Perspective

The Sun and Moon rotate in a circular pattern above the Flat Earth to create day/night cycles and seasons. The ancient Chinese Yin-Yang symbol is sometimes theorized to reflect the Sun and Moon over the Flat Earth. What causes them (and other celestial bodies) to move is less well understood, but magnetism, electric forces, and the aether are important considerations.

Some wonder if the Yin-Yang symbol represents the Sun and Moon over the Flat Earth.

In this framework, both the Sun and Moon are much, much closer than they are believed to be in the heliocentric Globe model. The Globe model says that the Sun is roughly 93 million miles away and the Moon is 238,000 miles away. Under the Globe model, the Sun is roughly 400 times bigger than the Moon, whereas a Flat Earth framework typically considers them to be much smaller and similarly sized (if not the same size). In fact, we can see the Moon's surface—with reasonable granularity—with our eyes from Earth. Some Globe Skeptics suggest this is evidence that it's much closer than 238,000 miles away.

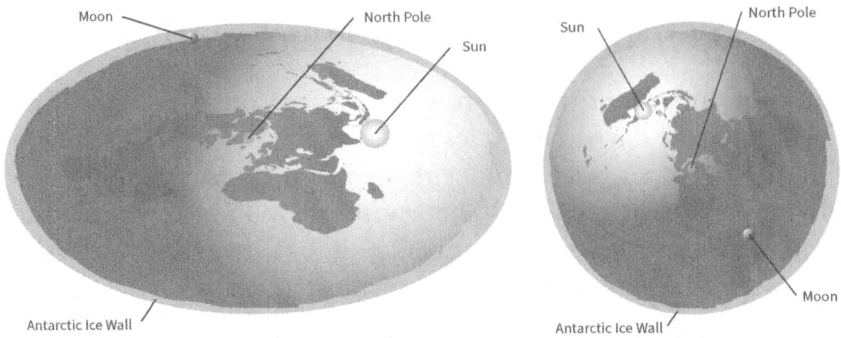

A visual representation of the Sun and Moon moving in a circular pattern above the Flat Earth, which generates day/night cycles and seasons. The visual is intended to be illustrative only and should not be regarded as precise.

Day and night occur on the Flat Earth because we cannot see forever, and light fades (known as *attenuation*). In other words, when light passes through a medium such as air, its intensity decreases. This phenomenon is apparent when shining a flashlight on a foggy day.[11] Because we cannot see forever, objects can seem to "disappear" far away, but they don't *really* disappear. They just move beyond what our eyes can see. So in a Flat Earth framework, it is indeed *expected* that not everyone would see the Sun all the time. It's also expected that the Sun would be visible to some and not others depending on its distance from the viewer and the local conditions. It's also noteworthy that the English language uses the words *sunrise* and *sunset* to describe the Sun: Our language builds in the *Sun's motion* that we perceive. Our language aligns with our senses.[12]

Additionally, *perspective* is essential to understand here: It refers to the way in which we perceive and interpret the world around us through vision. Artists and architects are well aware of perspective, but many others aren't, and a shockingly small portion of cosmological leaders talk about it.

Here's a basic example: Consider looking into the distance at railroad tracks, from the ground level, while sitting in the middle of the tracks. The tracks are parallel. Parallel lines, by definition, never intersect. However, as shown in the image that follows, the parallel tracks *appear* to converge in the distance. They are geometrically parallel, but to our eyes they look angled the farther away they are. They *look like* they will intersect because of the way we see.

Furthermore, far in the distance, the tracks are not visible to the naked eye: the tracks get smaller and smaller until they disappear, known as the "vanishing point." And that vanishing point looks like it is higher than the point of observation (the "eye level"). The eye level is where the horizon appears to be in the distance. As the distance from the observer increases, the tracks appear to be higher and higher above the viewer, and they appear to "rise" to the eye-level "horizon." They aren't actually rising. That's just how we see them.

Because of the phenomenon of perspective, parallel railroad tracks look like they are converging in the distance toward a vanishing point. When viewed from the ground, they appear to be rising.

Similarly, objects above us, like streetlights on the side of a straight road, appear to move *downward* as they are farther away from our physical location. They appear to be smaller and smaller, until they reach a vanishing point relative to the observer's eye level [see the image that follows].

Streetlights on a straight road appear to be dropping due to perspective, even though they are the same height.

Thus, objects overhead that move away from us can appear to be *dropping*, when in fact that appearance is a mere artifact of visual perspective. They aren't actually dropping; they're just moving farther away while staying at the same height above, and our perceptual system "interprets" this as "dropping." Unless we make this adjustment in our minds, we risk misinterpreting visual phenomena above us.

Thus, perspective explains the sunset from a Flat Earth perspective. The Sun doesn't set beneath Earth's curve in this framework, but instead it reaches a *vanishing* point in our perspective. In this regard, without adjusting for perspective, one might even erroneously conclude that clouds far away are "dropping below Earth's curve" too.

There are also documented instances in which the Sun can be seen to set *above the horizon*.[13] Globe Skeptics consider this to be one of many debunkings of the Globe because, according to the Globe model, the Sun should disappear *below the horizon*. However, Globe advocates typically claim that sunsets above the horizon are mere illusions caused by refraction (that is, the bending of light). (Note: Globe Skeptics often criticize Globe defenders for their frequent invocation of refraction as a post hoc rationalization whenever light-related phenomena can't be explained by the Globe.)

The Firmament

In the Flat Earth framework, the Sun's light shines through a container that we don't see with our eyes, often called a *firmament*. Typically, it is depicted as a dome that encloses Earth. The idea is that we live within a closed system but don't realize it.

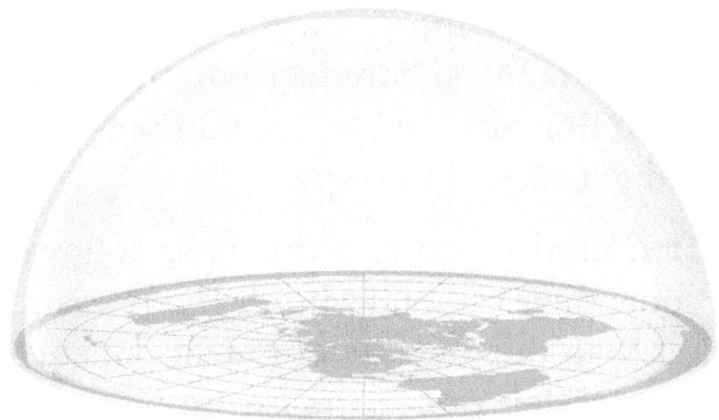

Above is a basic depiction of a firmament enclosure above the Flat Earth.

Many aspects of the firmament enclosure are unknown, such as what's beyond it, what exactly it's made of, whether it's physical or "energetic," and whether its shape is precisely a dome or something else. Regardless, as the theory goes, the firmament is something we cannot physically escape.

Steven A. Young supports this basic idea using his background in theoretical physics:

> Perhaps the strongest argument in favour of a stationary enclosed realm is the issue of air pressure. We live in a pressurised air environment, and it's fair to say that pressurised air requires containment. The Air with its barometric pressure of 14.7 pounds per square inch must

> be enclosed in a sealed container. When no air or water can get in or out, this is called Hermetically Sealed.
>
> The globe model claims that the surface of "planet earth" is wide open to the cold vacuum of space, but if this were the case, the atmosphere would get sucked off the planet instantly to equalise pressure with the surrounding vacuum (which is also an infinitely larger space). You can't have stable pressurised atmosphere that is not contained; gasses will always disperse into an area of low pressure. There are people who try to claim that "gravity is the container," that it somehow holds the gasses onto the outside of the planet in a "gravity well" in spite of the infinite vacuum all around; this is pure desperation. The theoretical notion of the "ionosphere" is also not sufficient to contain pressurised Air; it too would get sucked into space.[14]

An analogy often used to explain this is as follows: When you open a can of soda, there's a loud sound. The pressure from inside the can is higher than the pressure of the air outside, so when the can opens, the gas rushes out of the can to reach an equilibrium (that is, the gas *diffuses* out).

Compare this to what the Globe model posits. It suggests that Earth is not an enclosed system, and all of Earth's gases can sit peacefully next to an extremely low-pressure system (the "vacuum of space"). Somehow, Earth's gases don't violently diffuse into the low-pressure vacuum. The Globe's claim is that gravity holds in the atmospheric gases against this pressure gradient. This allegedly occurs at the highest altitudes when the "force of gravity" would be weakest because it's far from Earth's center of mass. Yet that force holds in the atmosphere against the immense pressure differential created by the massive (alleged) vacuum of space. Globe Skeptics don't buy this argument.[15]

Often Globe Skeptics point out that gravity is supposedly strong enough to keep Earth's oceans from falling off the Globe, but at the same time, gravity is not strong enough to hold down a bug that flies upward against the force of gravity. And yet despite all of this, Globe advocates claim that gravity is strong enough to hold in Earth's atmosphere against the pressure differential created by the vacuum of space that sits right next to the atmosphere. For Globe Skeptics, a firmament enclosure makes much more sense.[16]

Interestingly, a CIA document declassified in 2000 includes a dissertation defended in the USSR, which mentioned an investigation using "a visual photometer of the daytime sky intended for measuring the brightness of the firmament." It also discussed its formula "derived on the assumption of a 'flat' Earth."[17] Another Soviet scientific document, declassified by the CIA, discussed an invention by Moscow State University that "photographs the whole firmament at the time on a sensitive motion-picture film." It is titled "Scientific Station at Bukhta Tiksi Built Especially for Auroral Studies" (1959).[18]

Whether "firmament" has the same meaning as the Flat Earth interpretation of an enclosure, or if it's just a general statement of Earth's atmosphere, is open for debate. However, in the second declassified CIA document mentioned, the quotation uses the term *whole firmament*. If *firmament* meant "atmosphere," measuring the whole thing wouldn't make sense from one location on a Globe. However, the document's title specifies that the station was one location—Bukhta Tiksi, which is in the Arctic Circle. That would be near the center of the Flat Earth, and measuring the whole of the upper regions would be more plausible in that context.

The Lights in the Sky

Within the context of a firmament-enclosed Flat Earth, the lights in the sky (such as the Sun, Moon, stars, and planets) are much smaller than is assumed in the Globe model. And they do not exist in "the vacuum of space." Instead, they exist within some unknown medium above us—perhaps a type of fluid. There are debates about whether celestial bodies are *within* different layers of the firmament or beyond the firmament, but in all cases, the lights in the sky are much, much closer than they are in the Globe model. **For example, from the Flat Earth perspective, stars don't look like they're light years away...because they aren't.**

Planets, from a Flat Earth lens, are distinct from Earth. In fact, the word *planet* is derived from the Greek word *planetos*, which means "wandering star."[19] Objectively, that's how planets appear in the sky—their motion wanders relative to the other lights in the sky. The Flat Earth view is that Earth is not a planet but rather a plane (they are the same word minus the "t").

Seasons and the Analemma

The Sun's rotation above the Flat Earth moves closer to and farther from the North Pole center point. Its circular pattern closest to the North Pole is the Tropic of Cancer. That's summer for the Northern Hemisphere. The Sun's circular motion trends away from the Northern Hemisphere to the bigger circle reflected as the equator, and it goes even farther away to the biggest circle reflected as the Tropic of Capricorn (that is, summer in the Southern Hemisphere). This explains seasons on a Flat Earth. (Note: The word *hemisphere* is a misnomer when applied to a Flat Earth, because it references a *sphere*. But the word can be used to describe what's "above" and "below" the circle of the equator on the Flat Earth. The same goes for *atmosphere*, which is sometimes alternatively called *atmos*.)

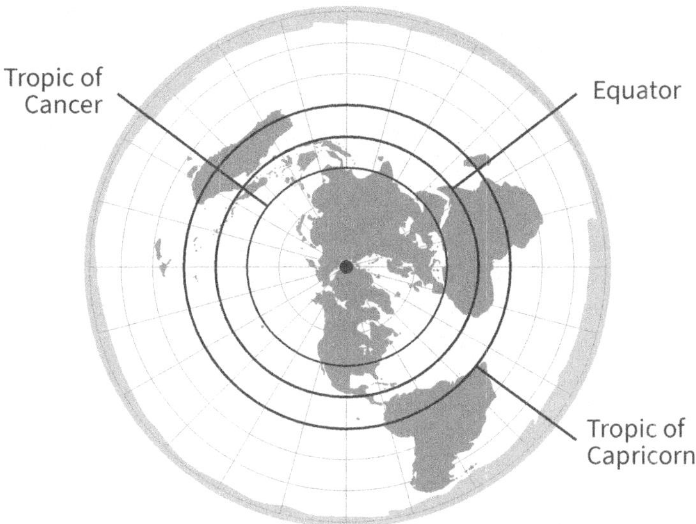

In the basic Flat Earth framework, the Sun rotates above Earth in circular patterns that are closer to and farther from the North Pole, which correspond to seasons in the Northern and Southern "Hemispheres."

The Sun has more ground to cover when it's closer to the Tropic of Capricorn during the Southern Hemisphere's summer. In other words, it's still making a daily rotation but has a larger circumference to travel over the same period—since one day is twenty-four hours. Thus, the Sun necessarily travels faster in its circular path during Southern Hemispheric summers. This aligns with the pattern of the Sun's location in the sky throughout the year, known as the *analemma*. This shape is found by tracking the Sun's position at the same time of day, every day. It forms an asymmetrical figure-eight shape (as shown in the image that follows).

The image above illustrates the analemma shape. It reflects a visual representation of the Sun's position in the sky when viewed from the same place every day at the same time throughout the year. In a Flat Earth context, the asymmetry of the shape aligns with the Sun's faster speed in the Southern Hemispheric summer because it has more distance to travel in a twenty-four-hour day (since its circular path is long), whereas, in the Northern Hemispheric summers, the Sun travels a shorter distance in a twenty-four-hour day and therefore moves more slowly.

The shape's asymmetry makes sense if the Sun is traveling faster and slower at different times of the year, as described within a Flat Earth. According to Globe Skeptics, the Sun's changing speed is reflected by changes in the amounts of twilight (the period before sunrise and after sunset during which the sky is partially illuminated).[20]

The Twenty-Four-Hour Summer Sun

This brings us to one of the biggest controversies in the Flat Earth discussion: the twenty-four-hour summer Sun in Antarctica. On the other hand, the twenty-four-hour summer Sun near the North Pole can be explained easily by the Flat Earth. The Sun is simply traveling slowly around a small circle near the center of the Flat Earth.

But how would the Flat Earth explain a twenty-four-hour summer Sun in Antarctica? The circle traveled by the Sun is much larger in the Southern Hemisphere than it is in the Northern Hemisphere. Because of this larger distance traveled, one would think that there would be periods when the Sun is on the other side of the Flat Earth, which would create darkness for those in Antarctica far away from the Sun. Globe advocates argue that observing the twenty-four-hour summer Sun in Antarctica would mark the end of Flat Earth, and Flat Earth advocates typically agree (unless there's some other mysterious variable that hasn't been considered).

The caveat, however, is that *the Sun itself* would need to be visible for the full twenty-four hours in order to contradict the Flat Earth framework as it is currently conceived. It's possible to have *light without a visible Sun*.

Due to a well-known phenomenon in physics called *coffee cup caustic*, the entire perimeter around the firmament-enclosed Earth could be lit up for twenty-four hours even if the Sun itself is not visible at all times (a visual depiction follows). So when the Sun is on one side of the Flat Earth, sun*light* could be seen by an observer on the opposite side during the Antarctic summer. That's the key distinction: sunlight versus seeing the actual Sun in the sky. Simply put: twenty-four-hour *sunlight* is explainable on a Flat Earth, whereas twenty-four-hour Sun is most likely not. In fact, twenty-four-hour sunlight, in which the Sun is not visible for the entire time, would seemingly debunk the Globe. The Globe predicts twenty-four-hour *Sun* visibility.

As discussed in Austin Whitsitt's June 2024 debate with "Professor Dave," the Flat Earth framework involves a phenomenon of light known as *coffee cup caustic*. When light shines onto a dome-like structure, the light wraps around. From this lens, twenty-four-hour sunlight would be possible in the Antarctica summer when the Sun is located on the other side of the circle of the Flat Earth—because the light wraps around the circular perimeter even though the Sun itself is far away.[21] The above images are an artist's rendition of the images shown in the debate and are intended to be illustrative only.

Globe Skeptics currently claim that although there are many films of the twenty-four-hour Sun in the Arctic region—a topic about which there is no debate—there are very few taken of the Antarctic region. And the few videos that have been released of the alleged Antarctic twenty-four-hour summer Sun have anomalies whereby it's not clear that the film is continuous. Or the film allegedly looks like it has been edited. Globe Skeptics assert that editing has even been admitted in some cases, which then puts all other videos into question. **Moreover, they note that governments show livestreams of Antarctica, but the feed apparently cuts out for hours every day, so there isn't *continuous* twenty-four-hour footage.[22]**

The debate could be resolved if Globe Skeptics and Globe proponents all went to Antarctica in the summer and filmed the sky for twenty-four hours. Plans have been discussed, but making it work can be challenging logistically and financially, not to mention governmental regulations that restrict independent travel below the 60th S parallel without permission (more on this soon).

Trusted parties on both sides of the debate would need to be present, and many replications would be required to provide convincing evidence. As contentious as this issue is, people probably won't believe what they see via a third-party video unless it aligns with their worldview.

Celestial Observations and the Azimuthal Grid of Vision

Within the Flat Earth framework, the stars in the sky rotate around the North Star, Polaris. Polaris stays above the North Pole (the center of the Flat Earth). People at different positions on Earth will see the sky, and the movement of the stars, differently.

This can be explained again by perspective—which requires a rewiring of how we interpret what we see. Recall that objects above us will look like they're moving down toward eye level, as they become more distant, when in reality they aren't (like streetlights in the distance on the side of a road). When perspective is applied to our entire visual space, it translates into a personalized 360-degree *dome* through which we see—known as an *azimuthal grid of vision*. It is dome-shaped because celestial objects above us look lower as they become more distant, whereas the highest point (as we perceive it) is directly above us.

This means that the way we look at all celestial phenomena needs to be recalibrated to account for the way we process visual information. Put another way, our vision is not linear (that is, Euclidean) but rather it's curved (non-Euclidean). **Thus, we have a curved limit to our visual space.** Moreover, we also have curved retinas. Curvature is simply part of the way we interpret the world through our perceptual system. What follows is a visual depiction of the Globe model's explanation.

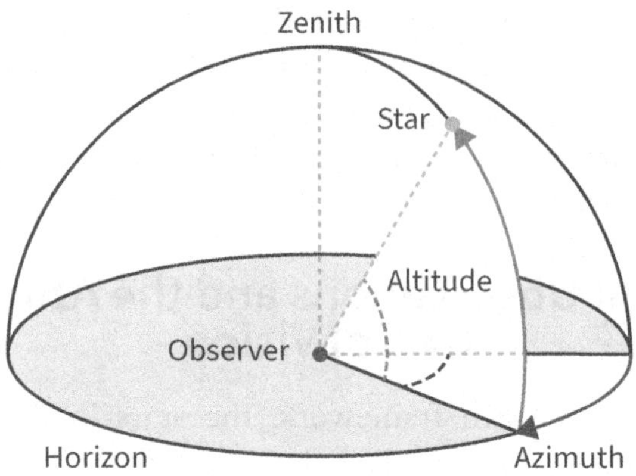

The illustration above shows the azimuthal grid of vision whereby we see in curved visual space due to the phenomenon of perspective. Objects in the sky that are farther away appear to be lower, thus creating the appearance that there is a 360-degree dome around us. The illustration has been re-created from Austin Whitsitt's 2023 presentation titled "Schooling Globers–Episode 7–Azimuthal Grid of Vision."

The personalized azimuthal grid of vision is responsible for various optical effects (such as optical convergence). These effects can be exaggerated because the atmosphere acts almost like a magnifying glass in the way that it distorts light. So what we think we see has to be adjusted in a big way.[23]

However, it's important to note that our curved vision is distinct from the firmament dome that is believed to be fixed around the Flat Earth. The azimuthal grid of vision is something each of us has, individually. Therefore, the sky takes on a different appearance *depending on our physical location on Earth*—since lights in the sky look the way they do because of their position within this personalized visual dome. On the other hand, the firmament is the same enclosure above and around all people on the Flat Earth.

It's often difficult to describe in writing the ways in which the azimuthal grid of vision impacts our view of the cosmos. The concept is probably best absorbed by watching videos, and a helpful primer is Austin Whitsitt's nearly three-hour 2023 presentation on YouTube, titled "Schooling Globers – Episode 7 – Azimuthal

Grid of Vision."[24] In any event, what follows are some examples in written form.

For instance, consider the phenomenon of crepuscular (and anti-crepuscular) sun rays. This occurs when the Sun can be seen in one location near the horizon, and at the same time, light can be seen 180 degrees away. So if you turn around, it *looks like* the Sun is at the horizon behind you as well as in front of you. In between, the sunlight looks like it is curving overhead even though there is not an actual physical bending of light. This phenomenon can be explained by our curved visual space.

Similarly, circular "halos" are sometimes seen around the Sun. The Globe model claims that this is caused by ice crystals in the sky. The same explanation is given by Globe advocates for "Moon dogs," which is a similar, halo-shaped phenomenon around the Moon. But when considering the personalized-azimuthal-grid-of-vision framework, no such ice crystals need to be invoked. The circular shapes we see as halos are simply aspects of our curved visual space. This can also be used to explain why rainbows look curved.[25]

Now consider a different and more basic phenomenon of perspective that relates to the way we perceive the stars in the sky. The stars appear to move counterclockwise around Polaris in the Northern Hemisphere and clockwise in the Southern Hemisphere. How could this be explained on a Flat Earth?

First, it's important to recall that stars in the sky will appear to be moving down, in the front of our visual space, as they are more distant. Whitsitt then applies the following analogy: Imagine a back-porch window that gets foggy, and you use your finger to draw a circle. You then keep tracing the circle in a clockwise direction. If someone looks at you from the other side of the window, it will look like you're tracing the circle in a *counterclockwise* motion in front of them. Thus, the direction of viewing determines the way the motion appears to you visually.[26]

The overarching point from these few examples is that apparently paradoxical observations can be resolved with an understanding of how we see. As Whitsitt says, the personalized azimuthal grid of vision allows us to "explain all celestial phenomena" within a Flat Earth framework.[27]

Aether Cosmology

Now that we've covered the basics, it's important to acknowledge an expanded version of the Flat Earth framework that is growing in popularity. It views the Flat Earth as part of a toroidal field with the North Pole at the center. The term *toroidal* refers to the *torus* shape, which resembles a doughnut. It is found throughout nature.

In this emerging framework, a toroidal field is above and below the plane of Earth (using technical terminology, the plane serves as a *block domain wall*—that is, *an inertial plane*). This model is often endorsed by Whitsitt and his colleagues at the Aether Cosmology channel (https://aethercosmology.com/), and it is also discussed in Steven A. Young's book *A Fool's Wisdom*.[28]

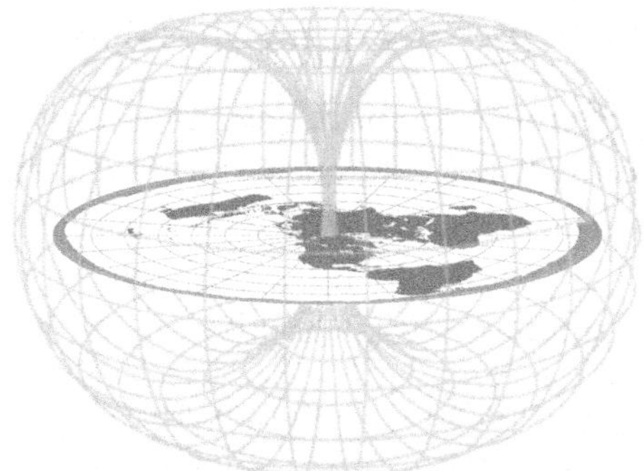

In the Aether Cosmology framework, the flat plane of Earth is within a toroidal field.

Restricted Explorations of Earth

While the discussed Flat Earth framework can explain many phenomena well, it has inherent limitations. This is due to restrictions in our ability to observe our home.

For example, consider that aside from airplanes, helicopters, hot-air balloons, and airships, human access to our upper regions of

Earth has been confined to a relatively small number of individuals, mostly from government-sponsored space agencies. Technological limitations also restrict our ability to survey the higher regions.

Imagery provided by third parties "from above" may be helpful, but Globe Skeptics often distrust what they're shown—especially when it comes from space agencies. As I'll discuss in chapter 6, the constant, admitted doctoring of images makes it rational for one to ask questions. Steven A. Young remarks, "Truly I tell you, all NASA footage is *unfalsifiable* and *unverifiable*; it's not scientific in any way at all, it proves *nothing*."[29] [italics in original]

However, there are some noteworthy government explorations of "what's above" that we don't hear the space agencies talking about regularly. In 1962, the United States detonated "nuclear weapons" in the upper atmosphere as part of Operations Fishbowl and Dominic. Many Globe Skeptics speculate that these were intended to test whether the firmament enclosing Earth could be penetrated. The name *Fishbowl* is curious in that regard, and *Dominic* is derived from a Latin word meaning "of/belonging to the Lord." Put together, these words lead to the phrase "Fishbowl of/belonging to the Lord."[30] This, in fact, aligns with biblical interpretations of the firmament. The Bible uses the term *firmament*, and many Globe Skeptics believe it to be a description of God's enclosure of Earth. The Bible also references "the waters under" and "the waters above" the firmament, which Globe Skeptics have interpreted to indicate that the celestial bodies above move within a fluid medium rather than an empty vacuum. As the interpretation goes, the firmament separates Earth and its oceans (the waters below) from the upper waters.

(Note: Whether the Bible supports the notion of a Flat Earth versus a Globe has been debated. One such debate was held between Globe Skeptic Rob Skiba and geocentrist Globe advocate Robert Sungenis in 2018.[31] Similarly, the Book of Enoch is sometimes invoked in support of a Flat Earth. It is a biblical text outside of the traditional canon, but the character Enoch was mentioned in the Bible. For more on this topic, see Zen Garcia's book *The Flat Earth: As Key to Decrypt the Book of Enoch* [2015].)

We also have limited knowledge of what's below us. **The deepest anyone has ever drilled into the ground beneath us is just under eight miles.** The expedition was the Kola Superdeep Borehole in Russia that was drilled between 1970 and 1994. For context, using modern geology's assumptions about layers in the ground, **this expedition made it about 0.3 percent of the way to the core of Earth**. And although we might want to believe that modern technologies can help us *infer* what's below us, the borehole failed precisely because of many unexpected findings—which means that our models were flawed.[32]

Geoscientist John Leeman, PhD, comments: "At this depth the scientists expected to find temperatures of about 100°C (212°F), but instead the ground was nearly twice as hot at 180°C (356°F). These extreme temperatures made the rocks behave more like a plastic material and eventually made drilling impossible." He further remarks on many "surprising" findings, including the types of rock they encountered and the prevalence of hydrogen gas and water.[33] So, who knows what's really below us? For that matter, as of 2023, only about 25 percent of Earth's ocean floor has been mapped out with high resolution.[34]

Finally, there are restrictions in our ability to explore portions of Earth itself. The Arctic region, for instance, has a military presence.[35] Although the area is theoretically accessible, there are restricted locations,[36] and certain areas are heavily monitored. In fact, the North Warning System (NWS) comprises radar stations established in 1988 (replacing the Distant Early Warning Line).[37] It is located at the 55th N latitude, from Alaska to Labrador, and is operated by the United States and Canada. We're told that it's for national security, scientific and environmental research, and air traffic control. Nonetheless, this represents government monitoring of a key area needed for an understanding of where we live. Similarly, the International Geophysical Year (1957–1958) established international collaboration to monitor the Arctic region (and the Antarctic region), involving nearly seventy countries. Again, on the surface, one of the primary objectives was to study "the environment."[38]

It's also worth noting that the North Pole has been considered an important location by many cultures. William F. Warren, a former president of Boston University, wrote about this in his nearly 500-page book published in 1885 titled *Paradise Found: The Cradle of the Human Race at the North Pole*, and similar ideas are discussed in *The Arctic Home in the Vedas* (1903) by Bal Gangadhar Tilak.

Immediately following the International Geophysical Year of 1957–1958 was the 1959 signing of the Antarctic Treaty. It restricts access to the Antarctic region past the 60th S latitude. Certainly people do travel to small portions of Antarctica today, but untold regions exist that have not been explored freely and privately. The penalties for traveling without a permit can be severe, and activities need to be preapproved. Antarctica is known as a protected environmental area (which Globe Skeptics take as a poor excuse to hide something), and there are even "environmental impact assessments." When one actually looks into what's required to explore Antarctica, it's very daunting.[39]

Austin Whitsitt provided an anecdote in 2023 in which he described reaching out to the authorities, informing them that he had a group of twenty people who wanted to freely and privately explore Antarctica. He offered to comply with regulations and requested a right to update where his group went once they saw what was there. But he didn't get any response.[40]

It's also worth noting that among the parties that signed the Antarctic Treaty of 1959 were the United States and the USSR. They were supposed to be enemies in a "Cold War," but they were apparently cordial enough to agree to restrict access to Antarctica. The treaty came after the United States Navy's "Operation High-jump" (1947–1948) and "Operation Deep Freeze" (1955–1956) involving the exploration of Antarctica.[41] **Upon returning from his expedition, US Admiral Richard Byrd called Antarctica "an untouched reservoir of natural resources" and spoke about a large amount of "unexplored" land.**[42] But given our current restrictions, such claims are difficult to verify. Commercial airlines don't fly over Antarctica, either.[43]

However, some individuals claimed to have found a passage within the "Antarctic ice wall" and traveled to "other civilizations." *The Iron Republic* by Richard James Morgan (1902) is one such case, though it's often debated whether the account is fictional. He wrote with apparent confidence that his tale validated a Flat Earth with lands beyond the ice wall: "It is unnecessary to say that after my narrative has been accepted, the…theory of Earth as a plane will have to be recognized and the geographies made in conformity therewith."[44]

Similarly, a book was published in 2022 titled *The Navigator Who Crossed the Ice Walls: Worlds Beyond Antarctica*, written by Nos Confunden. He shows a map of 178 worlds on our "flat" plane, of which Earth is just one. The other worlds exist beyond the Antarctic ice wall, and intelligent life exists in those locations. Stated another way, what we call "Earth" is just a single, small location within a much bigger landmass. Unfortunately, such claims will remain unverifiable until people are able to explore Antarctica without restriction.

A final point to consider regarding our inability to freely and privately explore Antarctica is airplane access. Both the Arctic and Antarctic regions have fewer airports than other places. However, a past form of air travel—the airship—might have enabled more freedom to travel, especially if technology had advanced to allow more forms of safe and independent landing.

As noted by Steven A. Young, the airship has been unavailable since the Hindenburg explosion in 1937. He speculates about whether there was an agenda behind the explosion itself. Given the panic around it, he wonders if it was planned to restrict travel. That is, perhaps the explosion was a "false flag" event. Many other tragic incidents have occurred with airplanes, cars, trains, and ships at sea, and those means of travel haven't been stopped. Yet the airship is no longer used.[45]

Ancient Beliefs about Earth's Shape

Although there are many restrictions that inhibit our ability to map out our home, cultures of antiquity all seemed to converge on a similar model: Flat Earth was a common belief. The fact that so many ancient civilizations believed in a Flat Earth could be viewed as compelling evidence for or against the idea. To those who believe we have "devolved" from past times, ancient beliefs in a Flat Earth would be viewed as a positive; and to those who believe in a more traditional evolutionary path, in which humans currently occupy the pinnacle of civilization, ancient cosmologies could be viewed as primitive and backward. (Note: Primitive or not, it's important to recognize that many ancient, Flat Earth–believing civilizations were able to accurately predict eclipses.[46] More on eclipses in the next chapter.)

In her book *Flat Earth: The History of an Infamous Idea* (2007), historian Christine Garwood elaborates on what earlier civilizations believed:

> Flat earth belief can be traced back to some of the most ancient civilizations in world history. The first of these are the Sumerians and Babylonians, who inhabited Mesopotamia, the land between the Tigris and the Euphrates (the site of modern Iraq) from c. 4500 to 500 BC. Although these peoples left texts describing a range of cosmological theories—too many to speak of a single, overarching Mesopotamian worldview—they developed the idea of a tripartite universe, with the earth as a flat surface ruled by the god Enlil, sandwiched between the sky and the underworld. For Egyptians, the same triple-decker arrangement applied, with the sky resting on four pillars, forked poles, or mountain peaks rising from the corners of the flat earth beneath....While later Egyptians made voyages to the so-called land of Punt, thought to be along the coast of East Africa, and other evidence suggests that they circumnavigated the continent, such experiences had no

> impact on ideas about the shape of the earth. This being the case, in the eyes of the oldest civilizations for which we have records, whether in the form of Babylonian clay tablets or Egyptian papyri, the earth continued to be a flat surface of a circular or rectangular shape.[47]

The ancient Hebrews also had a Flat Earth cosmology. Garwood asserts that the Old Testament itself "owes much to Mesopotamian mythology, which has led to claims that the Bible is a 'flat-earth book.'"[48] She further observes that Aristotle's (384–322 BC) writings suggest that his predecessors in Greek society believed in a Flat Earth.[49]

Additionally, some interpretations of Vedic (ancient Indian) scripture suggest a belief in a Flat Earth. As Radha Charma writes in his 2018 book *Vedic Universe: Flat Earth*: "What you will read in the pages of this book is Vedic wisdom that was once common knowledge, which has been covered over by the sands of time and unscrupulous personalities."[50] Many other ancient traditions believed in a Flat Earth as well, including ancient Chinese culture and the Mayans and Aztecs.[51]

The movement toward the belief in a spherical Earth can be traced to the teachings of Pythagoras (582–500 BC). While discerning the exact pattern of beliefs can be challenging, Garwood concludes: "All that can be said is that the Pythagoreans believed that the earth was a globe floating freely in space because the sphere was a perfect shape."[52] This philosophical and aesthetic bias toward spheres proved impactful and influenced others such as Plato and Aristotle who have had a dramatic impact on society since then. **Like the Copernican Principle—which is a philosophical bias toward believing "we are not special"—one might wonder if the somewhat arbitrary biases of the Greeks toward spheres has steered modern thinking without our conscious awareness. A philosophical preference is distinct from scientific evidence.**

A major landmark in the movement away from the Flat Earth cosmology is often attributed to the Greek scholar Eratosthenes.

In around 240 BC, he is said to have run a simple experiment that many today regard as early proof that Earth is a sphere. As written in Oxford University Press's *The History of Astronomy* (2003): "Ever since [Eratosthenes], everyone with a modicum of education has known that the Earth is spherical."[53] But Neil deGrasse Tyson explains that the results could be interpreted by a Flat Earth too. Tyson describes a portion of the ancient experiment:

> [There were] two cities in the old world and for one of them, they knew that at 12 noon on a particular day of the year that the Sun was directly overhead, and you can see the bottom of a well. How can we use this observation to see if Earth's surface is curved? We needed another [well]. Turns out we can't see the bottom of both wells at the same time. What might explain this? There are two possible explanations. First, we could have a flat earth with the Sun that's small and close by so that the light hits the second well at an angle. Or, second, we could have a curved earth with a Sun that's big and far away so that all the light comes in parallel, but only one well at a time is lit all the way to the bottom. **Turns out with just two wells there's enough wiggle room for both explanations to fit our observations.**[54] [emphasis added]

The observation that Tyson is referring to was part of the broader goal of Eratosthenes, which was to discover the circumference of Earth. The other part of the experiment involved comparing the shadows cast by sticks placed in the aforementioned locations. He arrived at a circumference value very similar to the one used today by the Globe model. However, his results with sticks and shadows could similarly be explained on a Flat Earth with a more local Sun. Eratosthenes simply presupposed a spherical Earth such that the numbers worked with circumference-calculation math. Interestingly, physicist Carlo Rovelli acknowledges that ancient Chinese explorers conducted an experiment in order to determine the distance to the Sun, with the assumption of a Flat Earth, and they arrived at a distance suggesting that the Sun is very close. Rovelli, a heliocentric Globe advocate, unsurprisingly calls this "the utterly wrong conclusion."[55] In any event, the point is that the story of Eratosthenes, contrary to popular belief, is insufficient to debunk a Flat Earth.[56]

More-Recent Globe Skepticism

Although the spherical Globe model has become overwhelmingly dominant in recent times—particularly in light of the Copernican bias—pockets of Globe Skeptics have persisted. In other words, questioning the Globe isn't a mentality that popped up out of nowhere.

Samuel Rowbotham's book *Zetetic Astronomy: Earth Not a Globe* (1865) sparked significant interest in Globe Skepticism in the 19th century. He described experiments suggestive of a Flat Earth, and he offered many of the arguments that persist in the growing movement today. William Carpenter's book *One Hundred Proofs That the Earth Is Not a Globe* (1885) advanced Rowbotham's ideas. Lady Elizabeth Blount founded the Universal Zetetic Society in 1892 and actively organized public lectures and debates on the subject. Wilbur Voliva implemented Flat Earth ideas into the Christian-dominated city and school system of Zion, Illinois, from 1906 until his death in 1942.[57]

In 1956, Samuel Shenton founded the International Flat Earth Research Society (IFERS) in Dover, England, garnering media attention and making television appearances. Charles Johnson took over IFERS in 1971 and moved the headquarters to California and continued to proliferate Flat Earth ideas. Separately, Leo Ferrari founded the Flat Earth Society of Canada in 1970 (later shortened to "The Flat Earth Society") and took a satirical and less serious approach than the concurrent IFERS headed by Johnson. The organization wound down in the 1980s, but Ferrari continued to speak loudly. Johnson resented Ferrari's approach and felt that he was undermining the seriousness of the movement.

As noted in Eric Dubay's "The History of Flat Earth" video, Ferrari would bring a large rock into interviews and claim he held on to it when his boat was falling off the edge of the Flat Earth.[58]

Recall that the notion of "falling off the edge" is not even possible within the predominant Flat Earth framework because it suggests that Earth is surrounded by an ice wall. Dubay views Ferrari's antics as a means of delegitimizing the broader movement by making a mockery out of it and discouraging people from actually looking into the arguments. Dubay even calls Ferrari's work "disinformation."[59] Johnson apparently called Ferrari "a false prophet, guilty of muddying the waters of truth."[60]

In 1995, six years before his death, Johnson's home burned down, and along with it, decades of research compiled by him and Shenton. Johnson felt that the fire was the result of arson by a NASA agent he believed had been seen "snooping around."[61] When Johnson died in 2001, that marked the end of IFERS.

In 2004, Daniel Shenton (no relation to IFERS founder, Samuel Shenton) created The Flat Earth Society, which remains in existence today. Kelly Weill, a journalist at *The Daily Beast*, notes in her book *Off the Edge* (2022) that Daniel Shenton has mainstream views on most things other than Flat Earth, such as believing the climate-change narrative and supporting modern medicine.[62]

Dubay sees The Flat Earth Society as a continuation of the "disinformation" from Ferrari because it makes claims that most modern Globe Skeptics do not believe—for instance, that the Flat Earth rises upward 9.8 m/s^2, forever, in order to account for "gravity." This is nothing like the predominant Globe Skeptic views discussed earlier. Thus, the Flat Earth Society is regarded among much of the Globe Skeptic community as a *psyop* (short for *psychological operation*, designed to misdirect and make a mockery of the subject—a "poisoning of the well").

The loudest segment of the Globe Skeptic movement is more in line with the serious research of Rowbotham and IFERS. In 2014, the ideas took off. This coincided with Eric Dubay's book *The Flat Earth Conspiracy*. It also followed the release of Robert Sungenis's 2014 documentary *The Principle*, which, as discussed earlier, promotes geocentrism—but not a Flat Earth. However, the notion of a stationary Earth in the center of the cosmos is an important part of the Globe Skeptic ethos. It's a gateway, of sorts, for Flat Earth beliefs. Sungenis, who believes that Earth is a sphere, suggests in his book *Flat Earth/Flat Wrong* (2018) that momentum around a Flat Earth serves to discredit the reality of a geocentric Globe.[63]

Also, video footage of "lunar waves," originally shot in 2012 by the host of *Crrow777 Radio*, might have contributed to the Globe Skeptic movement. The footage can be interpreted as showing an interference line—possibly a wave of fluid—moving through a telescopic video image of the Moon. The "wave" has been shot in front of Jupiter too.[64] Some believe that these observations support the biblical notion of "waters above": the idea that above the firmament enclosing Earth there is a fluid medium in which the celestial bodies reside.

As of the writing of this book in 2024, the Globe Skeptic movement—again, a movement that's distinct from the Flat Earth Society—is still growing. In fact, it's sometimes called *True Earth* rather than *Flat Earth*. There have even been "True Earther" online conferences since 2022.[65]

In a 2024 debate between Globe Skeptic Austin Whitsitt and Harrison Smith, a host of the popular alternative media organization *Infowars*, Smith noted that while he is not a Globe Skeptic, about half his audience is.[66] Also, in a 2024 interview, when National Football League (NFL) players Travis and Jason Kelce were asked how many NFL players believe in a Flat Earth, Travis replied: "Dude, honestly, there's at least 10 guys, 10 to 15 guys in every locker room I would imagine." Jason said, "You would be shocked....If you took an anonymous poll...like, nobody had to disclose their name or anything like that—you're over 15% of an NFL locker room. Maybe over 20."[67] According to a 2020 poll, roughly eleven *million* Brazilians believe that Earth is flat, which is about 11 percent of the population.[68]

The real figures are probably difficult to gauge, because in recent times, being associated with such beliefs has been regarded as shameful and idiotic. And yet there are many videos of pilots being asked about it, and they say that Earth is flat.[69] In fact, I had my own encounter (unexpectedly). At a luncheon while traveling abroad, I randomly sat next to a former pilot of more than two decades, and the topic of cosmology came up before I even knew he was a pilot. He said that he feels with near certainty that Earth is not a spinning globe, and that while most pilots hold on to mainstream Globe beliefs, some of them do share his views. He mentioned that the other Globe Skeptic pilots have to very quietly acknowledge each other because the belief is so heretical. He was also a teacher of other pilots and said he didn't teach anyone to adjust for Earth's curvature or motion.

Whitsitt makes similar comments about credentialed Globe Skeptics: "I know many people with PhDs and many credentials that do not believe the Earth is a spinning ball. Half of them can't even publicly say it because they would be fired or publicly humiliated and attacked."[70] He continues: "I think that if you have a profession where you teach astronomy or something like that, it would be super difficult for you to come out and say that everything that you've been learning and/or teaching other people was wrong. So it's expected that the majority of people think that the Earth's a globe. [The flat earth movement is] only growing. I actually know many professionals…every profession you can name—whether that be military, pilots, engineers, you name it—that think the Earth is a stationary plane—many of which cannot come out publicly."[71]

He also mentioned in a 2024 webinar: "There are multiple people that work at NASA that are Flat Earthers….I have a guy [I've been in touch with]—a former NASA employee stationed in Antarctica for six months; he said you never see the Sun for 24 hours. He was stationed [there] during their summer. He's a Flat Earther. There are people that are working at NASA actively right now that know that it's all a lie."[72]

Although unconfirmed cases like these are hard to verify, the statements are indeed noteworthy. And it would make sense for such individuals to typically stay under the radar.

However, sometimes public statements are made. In 2024, Lukáš Machala, a Slovakian Culture Ministry chief official, stated: "Has it been proved that the Earth is round? Have you been into outer space? No, nor have I, I don't know."[73]

Additionally, Polish astronaut Miroslaw Hermaszewski was asked in a 2018 interview: "You've been there! Is Earth really a sphere hanging in outer space?" He replied: "Earth is flat, as some expect. I didn't expect this question. I assure you: It's flat."[74] The alternative interpretation of these statements is that the comments have been taken out of context, and he was joking.[75]

Censoring Globe Skeptics

One of the hurdles for Globe Skeptics is censorship. "Flat Earth" searches on YouTube typically present a series of anti–Flat Earth material that alleges to debunk the idea while bolstering the Globe model. Often the debunking videos mischaracterize Globe Skeptics' positions or offer incomplete arguments. Unfortunately, someone new to this concept wouldn't know that without proper context.

Along these lines, during a Congressional hearing, Rep. Ted Deutch asked a YouTube representative, "What are you doing to combat conspiracy theories?" YouTube's Juniper Downs, Public Policy and Government Relations Global Head, replied: "The first thing is demoting low quality content and promoting more authoritative content. And the second is by providing more transparency for users. Such as searching for 'the earth is flat' on YouTube." Deutch replied, "Your response to that is to put a box saying, 'Nope, the Earth is not flat.'" She responded, "Correct."[76] One might wonder how YouTube determines "low quality" versus "authoritative" content, as this is a subjective distinction.

Artificial intelligence (AI) systems similarly reflect bias. This matter will likely evolve as the technology advances. Today, in 2024, when I asked Google's Gemini AI: "Is there any evidence that the Earth is not a spinning globe?", it responded: "There is a lot of evidence that the Earth is a spinning globe, but no credible evidence that it's not..." I likewise prompted: "Many people are presenting evidence that Earth is stationary and geocentric.

Is that credible?" It responded: "No, the idea that the Earth is stationary and the center of the universe [the geocentric model] is not considered credible by the scientific community. There's a wealth of evidence against it..."[77]

Finally, consider the drama that occurred with the 2018 *Netflix* documentary *Behind the Curve*, which covered the modern Globe Skeptic movement. Included in the documentary was an allegedly failed Flat Earth experiment run by Globe Skeptics. *Newsweek* then ran the following headline: "'Behind the Curve' Ending: Flat Earthers Disprove Themselves With Own Experiments in Netflix Documentary." However, the experimenters who were profiled in the documentary have since refuted this and have explained how the documentary misrepresented the study. They allege that they sent the producer information that clarified what they had measured, and that they did not debunk themselves, as it's been falsely claimed. However, that clarifying information was not included in the documentary. Thus, Globe Skeptics worry that the documentary misleads viewers.[78]

CHAPTER 5

GLOBE CHALLENGES

In the previous chapter, I set the stage for understanding alternative ideas about our home, whereas in this chapter, I focus more on direct challenges to the Globe. This approach is consistent with the theme of this book: trying to uncover what's *not* true in an effort to narrow down the possibilities of what *is* true. We know how flawed the current cosmological model is, so this exercise is necessary.

One way in which Globe Skeptics have approached this matter is to demonstrate that the assumed Globe radius value of 3,963 miles at the equator is inaccurate. If the radius value is wrong, then that has a ripple effect on the entire Globe cosmology, since everything depends on a specific size of the sphere—day/night cycles, seasons, and even the Globe Earth's relationship and distance to other celestial bodies (given the Globe model's assumptions about gravity, which relates to Earth's mass).

A number of examples in this chapter relate to such radius-value challenges. I address several other creative challenges as well (ones that don't address the radius value, per se).

A natural question often arises: *If we don't live on a Globe, then what is the nature of Earth?* As discussed in the previous chapter, one idea is that Earth is a flat plane that extends past the ice wall (or something like that). But if we escape the binary paradigm of "It's either a Globe or Flat," perhaps we'll open up to more creativity. What if our home is of a nature that we cannot currently conceive? Maybe parts of Earth aren't even purely physical and can't be perceived with our ordinary senses. Or perhaps the shape is not static, whereby the nature of Earth (and physical matter) can change over time.

However, even if we stick within a purely physical paradigm, for all we know, the shape could be a flat plane up to a certain distance past the ice wall; then it could curve downward as a sphere would—before becoming flat again and then stairstepping down to another flat plane. I write this partially in jest, but mostly it's intended to stimulate a thought process that tries to eliminate all preconceptions and "inside the box" thinking.

With this context in mind, I examine the following Globe-model challenges in this chapter. These are by no means the *only* challenges, but they are intended to provide a general sense of the types of arguments that are often being discussed:

- A Globe Skeptic's victory in court in 2019 (citing evidence of professionals who use flat, stationary-Earth assumptions; and examples of people who see things that should be below the curve on the Globe)
- Additional instances of "seeing too far" (examples not used in the court case)
- Ships disappearing "over the horizon" but brought back into sight with cameras
- The horizon rising to eye level as one's altitude increases
- Horizontal wave propagation over long distances
- Underwater sonar communication between whales
- Specular (mirrorlike) reflections
- Shooting stars, meteors, and asteroids

- Magnetic declination
- Time zones
- Selenelion (lunar) eclipses

If even a single phenomenon cannot be accommodated by the Globe model, then that model is, by definition, incorrect. Recall that a single black swan falsifies the model that "all swans are white." As the saying goes, "All you need is one."

A Globe Skeptic's Victory in a US Court (2019)

Globe Skeptic Zen Garcia won a case in 2019 in the Magistrate Court of Barrow County, Georgia, after running a contest asking for real-world proof of Earth's curvature or motion with two experiments. He said he would grant $5,000 to anyone who could do so. One would think that this would be a simple task. **But not a single person won the prize money.**

One man responded to Garcia with what he felt was proof. Garcia disagreed, and the man sued Garcia, asking for $15,000. The court ruled in Garcia's favor and did not grant the contestant any money. This certainly doesn't prove that the Globe has been debunked, but the fact that the judge sided with Garcia is noteworthy.

As Garcia recalled in a 2019 interview, the contestant who sued him provided "theoretical" examples that "were based on...his bias in the heliocentric worldview....I had written into the contest that this had to be based on real-world examples and had to be applied to something that could be measured and retested and determined by [other] people. And that in fact was not the case with what [the contestant] had created."[1]

As a part of his defense, Garcia submitted to the court numerous examples refuting the Globe model.[2] For instance, Garcia noted that surveyors, engineers, and architects don't factor into their projects the alleged curvature of Earth; and he mentioned canals, railways, and bridges that are constructed, sometimes over 100 miles, that are laid out horizontally—not accounting for curvature. Specifically, he included case studies on the Suez Canal (more than 100 miles long), the London and Northwestern Railway (roughly 180 miles long[3]), the Danyang-Kushan Grand Bridge in China (over 100 miles long), and others.

He also cited eyewitness testimony from the 1800s, including a surveyor who said in *Earth Review* magazine in 1896: "In leveling, I work from Ordinance marks, or canal levels, to get the height above sea level. The puzzle to me used to be, that over several miles each level was and is treated throughout its whole length as the same level from end to end; not the least allowance being made for curvature."[4]

Another surveyor wrote to the *Birmingham Weekly* in 1890: "I am thoroughly acquainted with the theory and practice of civil engineering....All our locomotives are designed to run on what may be regarded as true levels or flats....But anything approaching [the curvature of Earth] COULD NOT BE WORKED BY ANY ENGINE THAT WAS EVER YET CONSTRUCTED.... Horizontal curves on levels are dangerous enough, vertical curves would be a thousand times worse, and with our rolling stock constructed as of present, ***physically impossible***."[5] In an 1893 edition of *Earth Review*, an engineer commented: "As an engineer of many years standing, I saw that this absurd allowance [for Earth's curvature] is only permitted in school books. No engineer would dream of allowing anything of the kind."[6] [emphasis, including capitalization, in original]

Garcia did not mention in his court case the instances of more contemporary studies in which engineers conducted calculations without taking into account Earth's curvature. Consider the following examples:

- A 1953 military technical manual from the Defense Technical Information Center of the U.S. Department of Defense website, previously protected under US Espionage Law and titled "Atmospheric Refraction Errors for Optical Instrumentation," states in its introduction: "A comprehensive study of atmospheric refraction errors for optimal instrumentation, **based on a flat earth assumption**, will be published subsequently." The next section is titled: "**Validity of Flat Earth Assumption** for Atmospheric Calculations." It reads: "The relative mass of the atmosphere at any elevation angle is given approximately by the cosecant of the elevation angle. This relationship is correct for a flat earth and a flat atmosphere. **To obtain an approximate measure of the accuracy of the above flat earth relationship, it was compared with results obtained by a spherical earth equation.**" The analysis concludes, **"it appears that a flat earth assumption may be safely used."**[7] [emphasis added]

- A 1961 NASA paper titled "Calculation of Wind Compensation for Launching of Unguided Rockets" describes a trajectory simulation that "assumes a vehicle with six degrees of freedom and aerodynamic symmetry in roll and the missile position in space is computed **relative to a flat nonrotating earth**."[8] [emphasis added]

- A 1988 NASA Reference Publication titled "Derivation and Definition of a Linear Aircraft Model" says: "This report derives and defines a set of linearized system matrices for a rigid aircraft of constant mass, flying in a stationary atmosphere **over a flat, nonrotating earth**."[9] [emphasis added]

- A 2001 Army Research Laboratory paper is titled "Propagation of Electromagnetic Fields Over Flat Earth."[10] And another published in 2010, titled "Adding Liquid Payloads Effects to the 6-DOF Trajectory of Spinning Projectiles," states: "A 6-DOF rigid projectile model is employed to predict the dynamics of a projectile in flight. These equations **assume a flat Earth**."[11] [emphasis added]

- A 2003 Army Research Laboratory paper titled "An Energy Budget Model to Calculate the Low Atmosphere

Profiles of Effective Sound Speed at Night" states: "To briefly examine short range acoustic attenuation at night, we use the low atmosphere profiles of wind speed, temperature, and relative humidity...**as input to a flat earth**."[12] [emphasis added]

- The "Flight Dynamics Summary" document for aerospace students includes a section titled "making assumptions," which are used to "simplify." The first two read: "There is a **flat Earth**. (The Earth's curvature is zero.); There is a **non-rotating Earth**."[13] [emphasis in original]

- Bathymetry is the study of the floors of large bodies of water, such as ocean floors. In attempts to map out the topography of ocean floors, a ship sends sonar waves down to the ocean floor and measures the time it takes for the waves to bounce back. This process can involve the assumption of a flat sea level and a flat ocean plane as initial references points.[14]

Globe proponents often rebut these examples by arguing that such assumptions are made to simplify calculations. However, these are technical fields that often require precision.

The second line of evidence that Garcia references in the court case involves people seeing too far. In other words, they see things that should be blocked by Earth's curve. For a believer in the Globe, one would think it would be straightforward to find something in the distance that should not be visible because it is below Earth's curve. This simply requires having the appropriate camera or telescope. The physical curvature should block the object. Shouldn't there be a great abundance of these cases, well documented and replicated, to put the Flat Earth debate to an end?

But the reverse seems to happen repeatedly, which is perhaps one of the reasons why more and more people are expressing skepticism of the Globe. People buy high-resolution cameras, telescopes, or binoculars and see distant objects that should be physically blocked by the curve. The counterargument from Globe believers is that those things are only visible because the light refracts and creates a "mirage." Or they claim that the curvature math was incorrect or that the viewer's height was unclear (which would affect the viewable angle).

"Seeing too far" would inherently refute the radius value of Earth and thus debunk the Globe as it is currently conceived. But technically it would not totally debunk the concept of a spherical shape; it would simply debunk the idea that Earth is a sphere with a specific radius value. The shape could still be a sphere—just a sphere that has a much, much greater radius value than what the Globe model currently asserts. That would still reflect a radical departure from modern cosmological thinking because of the ripple effects on all other Globe assumptions discussed earlier.

Sometimes, however, no special equipment is needed, and people can still see too far (according to the Globe's assumed radius value). Garcia references cases in which people can see New York's Statue of Liberty on a clear day from up to sixty miles away, while it should be well below the curve. Similarly, he cites residents of Oahu, Hawaii, who can regularly see the island of Kauai from more than ninety miles away.

He also mentions one of the most famous cases that was picked up by South Bend, Indiana's ABC57 local news in 2015. The article, titled "Mirage of Chicago skyline seen from Michigan shoreline," reports: "A picture of the Chicago skyline taken almost 60 miles away, is actually a mirage. [Photographer] Joshua Nowicki snapped the pic Tuesday night from Grand Mere State Park in Stevensville. Under normal conditions, even when extremely clear, this should not be visible, due to the curvature of the earth. The Chicago skyline is physically below the horizon from that vantage point, but the image of the skyline can be seen above it."[15]

Garcia remarks in the court document:

> [The news report] was proven false, by two of my colleagues, Rob Skiba and Rick Hummer. These gentlemen went to Chicago and made a video documentary, confirming that one can...see the picturesque Chicago

> cityscape from the side of lake Michigan where Joshua took his picture. They then rode across Lake Michigan in a boat, while filming the entire duration, the city growing in scope and size as one came nearer to it. This documentary affirmed without a doubt that one could, in fact, see the Chicago skyline from such distance and that it was not as the [news'] weatherman claimed a mirage.
>
> He also interviewed several natives from that area, who all shared conclusively that seeing the Chicago skyline from such distance was a normal and daily possibility and that the only things which could skew in some slight manner such viewing, would be inclement weather, but that more or less it was...possible to see the entire cityscape even when in windy conditions and choppy waves from the opposite shores of Lake Michigan.[16]

Additional Instances of "Seeing Too Far"

Separate from Zen Garcia's court victory, many other examples have been observed and cited.[17] For instance, Austin Whitsitt has more recently referenced some striking examples in a 2024 presentation.

Before discussing the examples, however, it's first important to understand the notion of "the horizon." In the Globe worldview, the horizon is a physical location; it's where Earth curves *down and away* from the viewer and literally blocks the viewer's vision. What's blocked from view can change with altitude, but that doesn't change the fact that the Globe implies a physical horizon created by the curve. Since the Globe model assumes we're on a spherical object, we are always perceiving ourselves to be at the "top" of the sphere from our perspective, meaning that the sphere is always moving down and away relative to us.

On a Flat Earth, the horizon is only *apparent* based on a vanishing point with respect to one's visual "perspective" (as discussed in the previous chapter). The Flat Earth horizon actually moves around based on factors such as atmospheric conditions and perspective. This can be seen in time-lapse videos of the horizon itself; it moves up and down.[18]

In his presentation, Whitsitt provides a video example in which the horizon is observed to be much farther than it should be based on the Globe's assumed radius value. The video zooms in on oil rigs at sea, from the beach, and taken at a height of about a foot and a half off the ground. Whitsitt notes that the physical horizon should be under two miles from the camera at its height, based on the Globe's assumptions. The farthest oil rig is about ten miles away...and the horizon can be seen clearly *behind* it. The horizon "should be" about two miles away, but it's not. It is clearly behind something that's roughly ten miles away. This means that the horizon is much, much more distant than it should be based on the Globe model's math. **And since the horizon is a physical location in the Globe model, this should not be possible—unless the Globe model's radius value is simply incorrect. As Whitsitt says, "People will make all kinds of excuses, but it doesn't matter. It's basic geometry. The horizon [in the Globe model] has to be in front of the oil rigs. [There's] no magical invocation of refraction [that Globe-Earth believers can point to]. This right here refutes the Globe....This is why Flat-Earthers exist: they went and looked at the Earth."[19]** [emphasis added]

Whitsitt gives additional examples that he feels challenge the Globe's radius value. The examples, and Whitsitt's case as to why they're important, are as follows:

- Mountains 255 miles away from an observer at a height of 9,824 feet are seen, even though there should be 12,800 feet (more than two miles) of curvature blocking the view of the mountains. In other words, they should be *far* below the curve, and yet they are visible. The image is taken using infrared, which is significant because infrared *mitigates the effect of refraction of light*. Refraction is the standard Globe-based rebuttal. In other words, this observation is seemingly impossible on the Globe.[20]

- The longest line of sight observation in the *Guinness Book of World Records* is 275 miles away (taken in 2019 by Gaspard Picard in Mallemort, France). "Line of sight" refers to observations that are directly from the observer's eyes to the object that's viewed without any obstructions. In this example, the observer views mountains from an altitude of 9,251 feet (roughly 1.75 miles). The mountains should be more than three miles below Earth's curve, but they are visible.[21]

- The Canigou mountain in the South of France is observed, two days every year, with the sunset behind the mountain. In the video that Whitsitt shows, the mountain is not visible until the Sun sets behind it, effectively backlighting it so the mountain is clearly visible from long distances. The mountain should be far below Earth's curvature, according to the official Earth curve online calculator that Whitsitt used. The Globe advocate's standard rebuttal is that refraction makes the mountain visible. However, because the Sun is behind the mountain, what is seen is a silhouette. Whitsitt alleges that this effective "shadow" is the *absence of light*, and since refraction is the bending of light, refraction should not explain this observation. Certainly refraction is a continued point of contention, but the observation of the mountain alone is noteworthy. Whitsitt remarks that Globe advocates claim this is just an "illusion," and the mountain is not actually there. He calls this example "a direct falsification of the Globe."[22]

- "Mirror flashes" have been done over twenty or twenty-five miles. This involves placing a mirror on, say, a beach, and the mirror reflects light. Then the observer moves far away from the beach where there is a direct line of sight. The mirror, and the light it reflects, should be under a substantial amount of curvature at far distances. But that's not what is observed. The mirror flashes reach the observer at great distances. Whitsitt notes that the military has used this method because it's a silent form of communication.[23]

- In a separate presentation, Whitsitt references long-distance laser tests of more than thirty miles, over a large body of water. The laser is also viewed from the side that shows it's completely horizontal and parallel to the water—that is, the laser isn't refracting and coincidentally "curving over the Globe."[24]

The beauty of examples like these is that anyone can try to replicate them, and that goes for government agencies as well. Hopefully, over time, more clarity will develop as additional trials are conducted and replicated with ample documentation.

Comedian Owen Benjamin brings some levity to the situation with his cartoon skit called "Globe Earth Sniper." The sniper is hiding in the bushes, trying to shoot an enemy that he sees far away in the distance, but he constantly asks his compatriot in battle about adjusting for the Coriolis Effect (that is, Earth's rotation on the Globe). He thinks that the enemy he sees far away is just a mirage because he should be below Earth's curve.

His fellow sniper says: "We only have one shot at this guy [the enemy]. [Aim] a little left."

Globe Earth Sniper replies: "Okay, now factor in the spin [of Earth]....What direction are we? East, west, north, south, northeast? Factor in everything! We need to know exactly the direction, the spin of the Earth and how to combat it or else...."

His colleague urges him to just take the shot, to which Globe Earth Sniper responds: "But that would mean Neil deGrasse Tyson is wrong." His colleague urges him to shoot accurately because a miss would blow their cover on their covert location.

Globe Earth Sniper insists: "Tell me the spin of the Earth!" He then shoots and misses and says, "I guess [Earth] was spinning a little faster than I thought!"

He then says: "I can see the mirage," to which his colleague responds with urgency, "No, no, that's the target [the enemy you're supposed to shoot at]!"

Globe Earth Sniper insists: "It must be a mirage. I've done the calculation [of Earth's curvature]; I wouldn't be able to see him. It's a mirage. I'm not taking the shot." His colleague says: "It's him!" and urges him to just shoot.[25]

The skit continues, but the point is well made that when thinking practically about some of the Globe advocates' counterarguments, they seem like silly post hoc rationalizations. We experience the world one way and are told that it functions in another way.

Ships Disappearing "over the Horizon" but Brought Back into Sight with Cameras

One of Aristotle's famous "proofs" of a Globe, which is still used often today, was that when looking out at a large body of water, ships would disappear, hull before masts.[26]

Visual perspective can explain this without invoking the Globe model. Simply put, there is a vanishing point in our vision. But explaining why the hull (bottom) drops before masts (the tall part) requires a more technical analysis. The answer has to do with "angular resolution." The angle of vision determines what "disappears"—and lower objects will disappear first due to the viewing angle. The viewing angle thus creates the obstruction. Using more technical terminology from the field of optics, at a certain distance, an object reaches the "Rayleigh criterion" (that is, the diffraction limit). This hinders the ability to resolve details, and it's most apparent at low viewing angles. Thus, objects disappear "bottom up."[27]

However, Aristotle didn't have cameras like the Nikon 900, as we do today. So, many Globe Skeptics use these high-resolution cameras to zoom in when a boat "disappears" over the horizon. Thus, the camera alters the angular resolution, and the boat comes back into view. This implies that the boat is still on a level surface;

it's simply beyond view from the naked eye. This is said to occur repeatedly, even when the distant object should be "beneath the curve." Many such videos are available online. They show that the ship goes in and out of view with high-resolution cameras (see David Weiss's Flat Earth Sun, Moon & Zodiac Clock app, under the Frequently Asked Questions "Ships over the Horizon" playlist).

This phenomenon is another one that could use rigorous documentation and replication. But even if cases were published, one might wonder if any scientific journals would be willing to publish such heresy in a Globe-dominated world. As noted by geocentric physicist Robert Bennett, PhD: "Peer review is now peer censorship."[28]

The Horizon Rises to Eye Level

The Globe model would predict that as a person rises in altitude, the curve of the physical horizon should be more and more visible. So Earth would appear to be bending down and away, in all directions, and more so with increased elevation. This is simply a matter of geometry; the curvature of Earth determines all of this since the horizon is a physical location within the Globe model. However, that's not what happens. The horizon remains at eye level as one rises in altitude. Anyone can experience this when ascending on an airplane, for example.

This phenomenon is supportive of the Globe Skeptics' view, which suggests that the horizon isn't a physical location but instead reflects the vanishing point based on perspective. In fact, a seminal book for artists published in 1939, titled *Perspective Made Easy* by Ernest Norling, gives the following instruction for artists: "If we ascend in an airplane we shall find that the distant horizon rises with our height. It appears to remain at eye-level. This accounts for the peculiar basin-like appearance of the earth when viewed from a great height."[29]

Norling calls this "peculiar" because he assumes the Globe model is correct. But if that model is false, then the observations make sense.

Steven A. Young, PhD, summarizes the situation well:

> **All that would be needed to prove the globe would be to film the horizon while going straight up in a balloon; it would start flat at eye level, then as you ascended, it would bend at the sides and curve down and away as you got higher up. But nobody has been able to reproduce this; even SpaceX, with all the dozens of rocket launches they claim to do, never produced this simple footage of the flat horizon continuously bending into a sphere; their cameras always point down at the rocket. We learn nothing about the cosmos from these experiments....There are several amateur high-altitude rockets that have shown [that] the horizon remains totally flat at eye level, and even one that appeared to get stuck in the firmament. Globe sticklers will claim that continuous raw amateur footage is fake, while holding strong that NASA's piece-wise TV footage is all legit.**[30] [emphasis added]

It's also worth mentioning two pre-NASA anecdotes. In 1931, scientist Auguste Piccard ascended to an altitude of 51,775 feet (almost ten miles) in a hydrogen balloon. He and his assistant, who traveled with him, reached the highest point recorded at that time. *Popular Science* magazine reported: "In the first half hour, the balloon shot upward nine miles. Through portholes, the observers saw the earth through copper-colored, then bluish, haze. **It seemed a flat disk with an upturned edge."**[31] In other words, he didn't report a "down and away" curve but rather the reverse. [emphasis added]

In 1933, the USSR broke Piccard's record by launching a balloon to an altitude of 60,095 feet. *Popular Science* magazine wrote about this one too: "Soaring in their airtight balloon gondola to a record-breaking height of 11.8 miles above the earth, the other day, three Russian aeronauts brought back the first scientific observations ever made at so great an altitude. Above their heads, the sky provided a striking spectacle; its color had turned a soft, deep violet, and almost devoid of the light-reflecting haze found at lower levels. **Looking down, they tried in vain to detect any curvature of the earth's horizon.**"[32] [emphasis added]

Horizontal Wave Propagation

In 1901, Guglielmo Marconi sent radio waves across the Atlantic Ocean. Radio waves are simply part of the spectrum of light, but they are not seen with the naked eye. If they are sent to a faraway location on a Globe, they should fly off Earth into space as Earth curves downward. They shouldn't magically follow the curvature. However, on a Flat Earth, the waves would be sent without worrying about any such obstruction from a curve.

Marconi's transatlantic radio wave transmission was 2,200 miles. Before he attempted this, people were skeptical that he'd be able to do it successfully because they assumed that Earth's curvature would block the transmission at around 200 miles. Yet he was successful on his first attempt.

The response was not "This disproves the Globe," however. Instead, scientists thought that there must be a layer in the atmosphere that reflected the radio waves back to Earth, over the curve, and to the recipient. And when the waves were reflected, they landed at the recipient's location. This atmospheric layer is now known as the *ionosphere*. Globe Skeptics view this to be a post hoc rationalization to preserve the Globe: scientists invented an ionosphere in response to Marconi's success. Otherwise, they would have had to confess that the Globe model is incorrect.

However, as noted by Austin Whitsitt, even if the ionosphere is accepted as legitimate, certain waves should go *through* the ionosphere and not reflect down to Earth if they employ high frequencies. Here, he's referring to frequencies higher than the ones used in Marconi's experiment.

Whitsitt cites many examples in which radio waves with such frequencies successfully reached recipients more than 2,000 miles away. In other words, these radio waves shouldn't have reflected off the alleged ionosphere because higher frequency waves were supposed to *penetrate* the ionosphere. They should have just flown off into space under the Globe model. And yet they did reach the recipient. The argument is as follows: This makes sense on a Flat Earth because it's simply horizontal wave propagation on a plane. And at the same time, this result could be said to debunk the Globe model.[33]

Underwater Sonar Communication between Whales

Whales communicate by emitting sonar waves that propagate through the water and reach other whales at great distances—upwards of 10,000 miles. On a Globe, one would think that the sonar waves would go from the whale and fly straight out of the water as the Earth curves down at greater distances. It's just like the radio-wave example discussed earlier in which the wave would be sent straight out horizontally and then fly off Earth, into space, if not for an alleged ionosphere to deflect it back. But there is no ionosphere in the water. "Refraction" is also much more difficult to invoke.

On a Flat Earth, long-distance whale communication is easy to explain—the wave propagates horizontally from one whale to the other. There's no curvature to worry about. But the Globe model's explanation is that there is a channel in the water through which the sonar waves travel. That channel curves around the Globe. To Globe Skeptics, this is another post hoc rationalization that uses mental gymnastics to preserve the Globe—another invention out of thin air.[34] Or is it an accurate scientific explanation? The debates will inevitably continue.

Specular (Mirrorlike) Reflections

There are many examples of large bodies of water that clearly reflect the scene around them like a mirror, known as *specular reflections*. The water literally behaves like a mirror in the purity of the reflection (see below for one example, and many others can be found on Aether Cosmology's website[35]). If the water had any curvature, there would be a "diffused reflection" that distorts the reflection, as mirrors are highly sensitive to angles. Think about a "funhouse" where the mirrors have curvature and make people look taller and shorter, their features are distorted, and so on. A flat surface is critical to give undistorted, specular reflections.

Aether Cosmology's website summarizes the implications: "A flat and level reflective surface is required for an accurate specular reflection, due to the fact that the angle of incidence is directly tied to the angle of reflection. This allows the image size and shape to remain consistent with the object, within the specular reflection. On the other hand, any reflective surface that exhibits convexity or concavity will greatly distort the image by magnifying, diminishing, and altering its orientation, due [to] the angle of incidence."[36]

Above is an example of a specular (mirrorlike) reflection, which Globe Skeptics argue requires a flat surface. A curved body of water would introduce a diffused (distorted) reflection. Many such examples are included on Aether Cosmology's website.

The discussion about specular reflection relates to another concept often talked about among Globe Skeptics. They say that "water finds its level" and "large bodies of water lie flat." When a container is filled with water, regardless of the container's shape, the water will develop a level surface.

Imagine a basin with a curved and even jagged bottom. If it's filled with water, the surface is still flat in spite of the nonuniformity below it. The exception would be a very small amount of water, like a drop, that has a spherical appearance.

So, from the Globe Skeptic perspective, the Globe itself is an anomaly because it asserts the existence of large, curved bodies of water that hug the sphere around all of its curves. Globe proponents explain this by claiming that water is held into place because of gravity. But as discussed previously, the notion of gravity (as classically defined) is entirely up for debate.

Shooting Stars, Meteors, and Asteroids

We see "shooting stars" fly across the sky above us as flashes of light. According to the Globe model, they are pieces of space debris that burn up in the atmosphere. On a Globe, however, the atmosphere curves all the way around the sphere. Space debris should exist around us in all directions—it should be above us, to our sides, and even directly below us, if we went far enough. That's because the Globe is a ball floating in space, surrounded by celestial bodies from all directions. When we look up, we're only seeing one part of space.

In theory, then, we'd expect space debris to come at us from all angles, including from below the Globe and *up from the horizon*. In fact, what one person observes as a shooting star above them would appear to a person on another part of the Globe as a shooting star coming up from the horizon (as one example). They'd see the same event from different angles. **Yet we do not see shooting stars come up from the horizon.** The same goes for "meteor showers": They aren't observed coming *up* from the horizon. They are always directionally "above us" in the sky.

Some Globe Skeptics view this as a debunking of the Globe model. On the contrary, this observed pattern of "shooting stars" would make sense on a Flat Earth in which we only see such activity in the sky above.[37]

The notion of space debris raises other questions that need to be addressed from a Globe Skeptic perspective. Lights in the sky that are identified as "space debris" could simply be electrical activity, they could be an excitation of the aether, they could be instances of government experimentation, or they could be something else not understood. We can't assume that they are physical objects unless they fall down to us. The notion that we are seeing physical rocks in the sky is ultimately an inference.

And even if there are physical rocks that fall from the sky and are collected, it's difficult to know with certainty what their origin is. For instance, how would we know that a rock definitively came from "outer space"? Flat Earth advocates might wonder if elements of the firmament could fall. Or, alternatively, perhaps rocks seeming to fall from the sky came from human technology. There are many causal explanations that would need to be explored.

Along these lines, the notion of "craters" created by "space rocks" is often questioned by Globe Skeptics. For example, what's believed to be a roughly 50,000-year-old crater in Arizona, with a nearly one-mile diameter, is alleged to have come from a meteor collision. And yet the large meteor that allegedly caused it is nowhere to be found there (see the image that follows). Globe Skeptics question whether an object so big should still be near the site of impact, to some significant degree. A rock so enormous would likely be difficult to move. Yet all that's seen is the crater. This pattern emerges with other crater shapes found in many geographic locations. Certainly scientists can collect small pieces of rocks nearby and claim that the origin is from outer space, but once again, determining the origin relies on inference. In the absence of large debris, scientists sometimes infer that the meteor must have largely "vaporized."[38]

The Barringer crater in Arizona is often assumed to have been created by a meteor. But are there other possible explanations for this geological formation?

In fact, invoking "rocks from space" is not necessary in order to explain the shapes identified as "impact craters." For instance, geysers that are not erupting have a similar configuration. Also, the 2022 PBS *NOVA* documentary *Arctic Sinkholes* explores geologic explosions—coming from *within Earth*—that result in crater-like appearances on Earth's surface. These structures pop up "out of nowhere." One of the scientists in the documentary comments that when she first heard about this, she "didn't believe it" and "thought it was a made-up story."[39]

The lack of skid marks near the crater impact is noteworthy because this seems to imply that the space debris comes down to Earth, every time, as if it were a vertical impact. But one might expect more angular variation, resulting in skidding and different types of divots in the ground. In the Globe model, there should be space debris all around the sphere, and the debris should be shooting toward Earth at *all angles*, not just from directly above. But instead, we see relatively uniform crater shapes that imply a vertical impact.[40]

Also, it's important to consider the possibility of government ballistics and explosions, and even firework technology, that might be erroneously interpreted as space debris.[41] Ultimately, many possibilities would need to be considered for all such phenomena before conclusions can be drawn.

Magnetic Declination

Magnetic declination refers to the difference between the Globe's geographic and magnetic poles. In other words, the geographic North Pole is not at the same location as the magnetic North Pole, and the same goes for the Globe's geographic and magnetic South Poles. A visual is shown below.

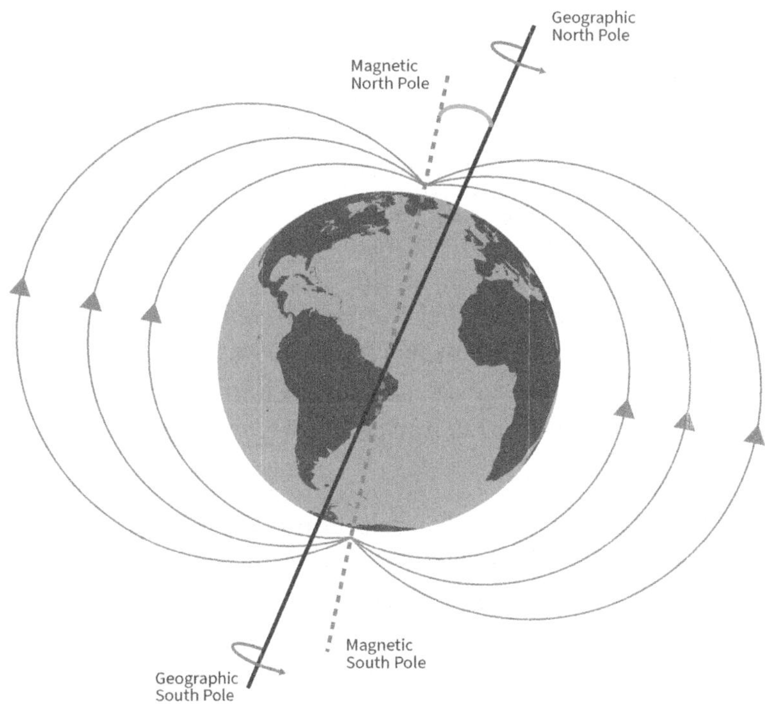

In the Globe model, Earth's magnetic field is generated by the core. The magnetic North and South Poles are in different locations than the geographic North and South Poles. The visual above is intended to be illustrative only.

This discrepancy exists because Earth's magnetic field is allegedly generated by a complex process inside Earth's core, which results in misalignment with Earth's "rotational axis." We've already discussed at length that the belief in Earth's motion is problematic, but assumptions about Earth's core deserve to be questioned too.

Recall from the previous chapter that we have a limited understanding of what is below us. The farthest down we've been able to dig is *less than eight miles* (the Kola Superdeep Borehole). The core is supposed to be *almost 4,000 miles down*. And the Kola project encountered many "anomalies" on the way down, indicating that geological models of the first eight miles were objectively not accurate. Why, then, should we believe in the accuracy of geological models telling us what is thousands of miles below us?

That leaves us in a concerning place: the belief that Earth's magnetic field comes from the core is ultimately theoretical.[42] More specifically, geophysicists tell us that Earth's magnetic field comes from the "outer core." The inner core, in their model, is too hot to maintain magnetic properties. That is, the temperature allegedly exceeds what's known as the *Curie point*. So, instead, geophysicists conclude that the spin of the outer core generates Earth's magnetic field, which is called the *geodynamo theory*. Researchers at the University of Maryland's Department of Physics wrote in 2020 about an ongoing project to try to demonstrate the effect, but confessed, "It's never been tried before, so the results are hard to predict."[43]

Furthermore, physicist R. Russell Humphreys, PhD, wrote an article in 2013 titled "Planetary Magnetic Dynamo Theories: A Century of Failure." He summarizes his findings: "After reviewing analytic theories, computer simulations, and laboratory experiments, I have concluded that all those efforts have fallen short of proving the geomagnetic field could be maintained by a dynamo."[44]

In other words, a fundamental aspect of the Globe model—the nature of the core itself—is not firmly established. Austin Whitsitt comments on the implications: "[The geodynamo theory] hasn't met the parameters of scientific verification....You have to be able to replicate it....You have to be able to show that it's viable.... [The Globe model] can't even demonstrate viability at the core of the belief."[45]

This shaky foundation should automatically raise questions about the Globe's magnetic declination—which, as a reminder, refers to the difference between geographic and magnetic poles. On a Flat Earth, however, there is no such discrepancy because the magnetic field comes from the geographic North Pole at the center of the circular lake containing Earth.[46] There is no magnetic declination problem on a Flat Earth. From a Flat Earth lens, "south" is simply anything directly away from that center point, so there isn't even a South Pole.

In the Globe model, not only are there discrepancies between geographic and magnetic poles, but the discrepancies are not symmetrical. Geographic and magnetic North Poles are relatively close to one another, while the magnetic South Pole is *much farther away* from the geographic South Pole. But if Earth is generally shaped like a sphere, one would expect something closer to symmetry, especially since the magnetic field is supposed to be generated from the *center* of earth (Earth's core).

Moreover, the National Geophysical Data Center shows that the magnetic South Pole has moved farther and farther away from the geographic South Pole over hundreds of years, whereas over that same period, the locations in the north haven't deviated quite as much. Ironically, as pointed out by Whitsitt, a representative of the National Geophysical Data Center acknowledges that the magnetic field is changing, but says that *why* this is happening is not understood. Therefore, it's difficult to predict what will happen a few years in the future. And in the same breath, the representative says they think there will be a "pole reversal"—which immediately contradicts the notion that it's difficult to predict Earth's magnetic field. This also raises questions as to the accuracy of their historical measurements of the magnetic field—which makes sense given the theoretical nature of the "core."[47]

But perhaps the problems discussed make sense when considering the possibility of a Flat Earth. Whitsitt explains that if one were to spread out a basic rendition of a Flat Earth over a Globe, one would need to significantly compress the Globe's south. The north would need to be stretched out, but only by a small amount.

As Whitsitt points out, magnetic "anomalies" might thus reflect the attempt of Globe advocates to slap the Flat Earth onto a sphere. This notion is supported by the fact that there is almost no magnetic declination around the Globe's equator. The equator also happens to be where the basic Flat Earth and Globe maps would be the same. The problems arise when the maps are most different, which is in the south. Adopting a basic Flat Earth map would alleviate these magnetic "anomalies."

Let's now consider this from a practical standpoint, because magnetic declination has real-world implications for navigation. Navigators are steered by a compass to geographic locations. Their compass steers them based on a *magnetic* direction. **But because of the discussed geographic and magnetic discrepancies (that is, declination), the Globe's navigation constantly requires "corrections" in order to reach the desired geographic location. Such corrections are laid out in magnetic-declination correction charts.** Navigators who don't make these adjustments risk getting lost because their compasses would lead them to incorrect geographic locations. To be clear: the Globe model forces us to manually correct compass readings in order to reach desired geographic locations.

The necessary corrections are much greater in the south than they are in the north, and the corrections are minimal at the equator. In fact, the corrections get wider and wider the farther south one travels. As pointed out by Whitsitt, at https://www.magnetic-declination.com/, magnetic declination data is freely available for all locations on the Globe. In the Antarctic region, the adjustments are almost hard to believe. **For instance, Latitude: 76° 29' 9.8" S; Longitude: 141° 38' 35" E has a declination reading of -178°. Essentially that instructs the navigator to turn around. This means that your compass will *completely* misdirect you without**

correction. Not far from that location is a *positive* declination of 117° (at Latitude: 72° 0' 46.6" S, Longitude: 164° 22' 38.9" E). These are just two examples out of *many* such discrepancies.[48]

More generally, at the 60th S latitude—which is where the Antarctic Treaty starts to restrict travel—the corrections go haywire. One might then speculate that restricted travel to this region allows the problems of the Globe's magnetic model to remain less obvious, thereby hiding the Flat Earth.

For instance, Whitsitt cites R.G.S. Collamore's book *His Pronouncement*, published in the early 1900s. The author quotes many individuals who reported consistent problems with navigation. As it pertains to the Southern Hemisphere in particular, he references a striking case in the report of *The Cruise of the Carnegie*: "We were astonished as the *Carnegie* proceeded south toward the region of Queen Mary Land to find **the chart errors in declination constantly increasing**, until, in the region of latitude 60° S longitude 110° E, they reached a maximum of 12° for the U.S. and British charts, and of 16° for the German charts."[49] Collamore then remarked: "It is a conspicuous fact that although these errors result in terrible disasters, there is little activity as to investigations, and there is an apparent attempt to discourage efforts to solve the mystery. Such a state of affairs naturally gives rise to such questions as: Why eleven to eighteen miles error in estimates every time? Why the apparent secrecy?"[50] [emphasis added]

Collamore then provided a September 20, 1862, quotation from *The Builder*: "Assuredly there are many shipwrecks from alleged errors of reckoning which may arise from a somewhat false idea of the general form and measurement of the earth's surface; such a subject, therefore, ought to be candidly and boldly discussed."[51]

In summary, as Whitsitt says, "You have to cook the books" in order to navigate on a Globe.[52] And given the problems discussed with the unvalidated model of Earth's core, perhaps this is not surprising.

Conversely, navigating on a Flat Earth is simple because geographic north is the same as magnetic north, and all navigation flows from that. Whitsitt further attributes all of this to an intentional desire to deceive by forces that might want us not to know where we live:

> This deception is pretty brilliant and they [that is, the conspirators] used it to convince people that navigate, both in a plane and on a boat, that they are traveling on a globe. Because when they use the globe assumption, they get to where they're supposed to go. This is because they couldn't actually use the globe to navigate. You literally cannot use the globe to navigate; you have to claim the compass doesn't work and then update and fill in the discrepancies from the globe model with magnetic declination....That's how they came up with it in the first place,...certainly in the South, when they really needed to save the globe. And now people navigate with the correction to save the globe, and they just don't even think about it. So they navigate assuming a sphere, making the corrections, and then they're like, "Oh yeah, it works."[53]

On a Flat Earth, the magnetic-declination problem disappears entirely.

Time Zones

In theory, time zones should be related to sunlight, which is objective and predictable. Since there are twenty-four hours in a day, twenty-four equal time zones would make sense. However, we currently have nineteen time zones in the Northern Hemisphere and thirty-two in the Southern Hemisphere. Globe supporters claim that this is solely due to political reasons, whereas Globe Skeptics see this as intentional deception that's part of hiding the nature of Earth.

Hervé Riboni was a professional sailor and a participant in the 1993–1994 Whitbread "Round the World" race. He is also a Globe Skeptic who created a presentation about the ways in which time zones show a deliberate concealment of a Flat Earth.

He provides his professional background and explains what led him to ask questions:

> As a professional sailor, I sailed ocean races, and I experienced sailing around the world along the 50th latitude south, which goes by the Cape of Good Hope at the southern tip of Africa, Cape Leeuwin in Australia, and Cape Horn at the bottom of South America. I have also crossed the Atlantic eight times and traveled more than 100,000 nautical miles.
>
> When I was a skipper, acting as captain, I frequently utilized maritime maps, and when I discovered the flat earth theory, I asked myself how the manipulators went from a flat surface to a sphere, without it being obvious that the sphere does not work when taking into account travel times, distances, and the positions of the continents. What method did the manipulators use to fool the entire world without it being obvious on the maps we use?[54]

Riboni discusses time zones as an important mechanism. First, he points to three missing hours in the north: while it's midnight in Russia, it's 3 a.m. in Alaska, and they are right next to one another. Westward, in the north, there are two hours of time zones missing as well. So, five time zones have been "hidden," leading to nineteen time zones in the north.

On a Flat Earth, the "northern" parts are closer to the center of the "lake" (the North Pole), which covers a much smaller area than what's farther south. On a Globe, however, the north and south are symmetrical. So the "stretching" of time zones in the north relative to the "compressing" of time zones in the south helps to account for this while maintaining a Globe. And yet at the equator, where the basic Flat Earth and Globe maps would match, there are "coincidentally" exactly twenty-four time zones.[55] Recall that magnetic declination is largely absent at the equator too.

Many of the "extra" time zones in the Southern Hemisphere are in the ocean, which makes their presence less noticeable. Riboni further remarks that "the international date line is the ideal place to add or delete time zones because the change of date makes the missing hours or extra hours much less obvious, especially given that the people on either side of the date line are not on the same days, making the comparison much more difficult."

Comparing the United States with Australia is a useful exercise because they have almost the same widths on the Globe map. Their positions are also similar enough in latitude relative to the equator to compare the time zones. In spite of their similarities, the US has four time zones and Australia has two (technically three, if the half-hour time zone is included in the count). Riboni comments: "If we superimpose their position and width on a globe, we see that the width of the time zones between the north and south relative to their latitudinal position are not enough to justify the difference of double the number of time zones....In the middle of Australia, they have added an extra half-hour time zone, to make us believe in a third time zone."[56]

In summary, Whitsitt describes the Globe Skeptics' view of time zones: "[It's a] total coincidence that it happened just in the same way you would need to use the time zones to make it look like it's a globe if it were flat. Just a coincidence that all the governments made these decisions. That's the actual argument [made by Globe believers]."[57]

Selenelion (Lunar) Eclipses

The final Globe challenge discussed in this chapter is particularly striking: the selenelion eclipse. This occurs when the Sun and the eclipsed Moon are visible...simultaneously. Lunar eclipses, from a Globe perspective, are supposed to occur when Earth blocks the Sun's light, casting a shadow on the Moon. But if both the Sun and the *eclipsed* Moon are visible at the same time, then Earth must not be blocking the Sun's rays. Globe Skeptics often view this as a clear debunking of the Globe model because it fundamentally violates their theory. As Sciencenotes.org states: "A selenelion [eclipse] seems impossible."[58]

However, the Globe rebuttal is "refraction." The argument is that light bends, so we're not really seeing the Sun and the eclipsed Moon at the same time. It's just a mirage. They're both below the horizon, and the light bends so you see them. This has to be the Globe argument because otherwise the phenomenon would be the end of the Globe. Globe Skeptics, however, contend that the "refraction" argument is no more than a post hoc rationalization. What follows is a visual depiction of the Globe model's explanation.

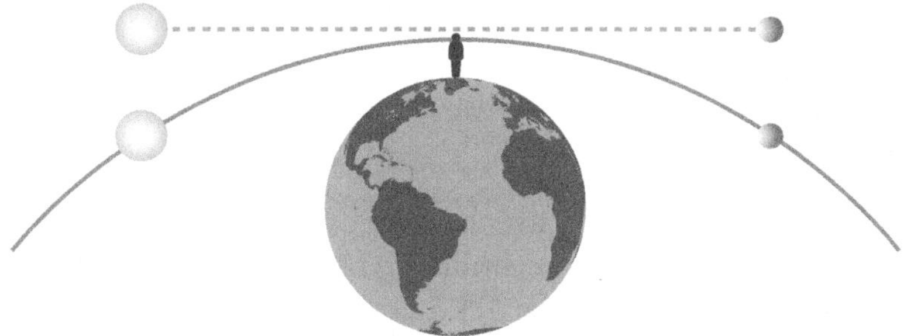

In a selenelion eclipse, the eclipsed Moon and the Sun are visible to an observer at the same time. This shouldn't be possible within the Globe model. The Globe model explains this, however, by alleging that people are seeing a mirage because light is refracting (bending). This hypothesis is required for the Globe model to remain intact.

The lunar eclipse does, however, raise questions on a Flat Earth too. Why does the Moon look like it's being obstructed? Various possibilities are proposed, but as of now, no one knows with certainty. (Note: See the video playlist presented in David Weiss's Flat Earth Sun, Moon, & Zodiac Clock app, under the "Eclipses" playlist within the Frequently Asked Questions.[59])

What we know is that *something* is happening whereby the eclipsed Moon is not fully illuminated. Either the light source is somehow "turned off"; or alternatively, something with a seemingly circular shape is getting in the way and blocking the light. The latter concept aligns with Vedic astrology; in fact, "Rahu" and "Ketu" are said to be bodies in the sky that cause eclipses. In Hindu mythology, Rahu is a demonic being that "swallows" the Sun and the Moon. Put another way, this idea suggests that celestial bodies exist in the sky that we don't see with our eyes, but we see their effects when they block light sources. (Note: One might even wonder if such invisible celestial bodies, and even visible ones, influence our world in ways not acknowledged by mainstream science. This would lend credence to astrology. The concepts presented here require a truly open mind and a "start from scratch" mentality.)

Along these lines, Crrow777's documentary *Shoot the Moon* (2019) includes telescopic video footage that appears to show a second Sun-looking body in the sky. After careful consideration, he doesn't feel that it is an optical effect because he did all he could to eliminate lens flare. Ancient mythologies have referred to multiple suns, including Chinese mythology, which talks about a period during which ten suns were in the sky.[60] Others talk about a "Black Sun" beneath Earth.[61]

Finally, it's worth considering the *solar* eclipse in this context too. This phenomenon occurs, according to the Globe model, when the Moon blocks the Sun's light. It's a convenient coincidence, however, that the Moon and Sun appear to be the same size during such eclipses: the Moon perfectly blocks the Sun. Yet, under the Globe model, the Sun is about 400 times larger and 400 times as far from Earth as the Moon. We are told that the Sun is 93 million miles away from Earth, and the Moon is 238,000 miles away; and the Sun's diameter is 864,000 miles while the Moon's is 2,159 miles. **Under the Globe model, the math just happens to work out perfectly such that the Sun and Moon look like they're the same size during the solar eclipse.**

An alternative explanation is that the Sun and Moon are about the same size rotating above the Flat Earth. And both are much, much closer to Earth than we're told within the Globe model—and much smaller. So, one explanation for the solar eclipse would be that the Moon is indeed moving in front of the Sun as they both revolve around the Flat Earth above, almost like the hands of a clock that overlap occasionally.[62]

Another possibility is that a mysterious celestial body moves in front of the Sun (such as "Rahu," as discussed on the topic of lunar eclipses). Or perhaps there is another explanation that hasn't yet been considered.

More broadly, the Moon itself is a mystery. For example, peer-reviewed research suggests that women's menstrual cycles can synchronize with the Moon.[63] What characteristics about the Moon, a celestial body in the sky, would impact human biology on the surface of Earth?

There are even studies suggesting that moonlight has strange characteristics. While sunlight creates warmer temperatures than those measured in the daytime shade, *moonlight creates colder temperatures than the nighttime shade.*[64] Globe Skeptics often speculate that the Moon is a self-luminous body. In other words, the Moon is lit up in the sky not because of a reflection from the Sun but rather because it is its own light source. Along these lines, some theorize that the Moon isn't a sphere but might simply be a round disk or some other shape that is interpreted to be a sphere from our vantage point. In fact, we see the same face of the Moon, always. (Note: This argument can be applied to any celestial body observed from Earth. Determining the true shape of anything would require 360-degree imagery.)

There are even historical reports of seeing stars through the part of the Moon that's not lit up.[65] This could be interpreted as evidence that the Moon is not fully solid. There are indeed many country flags that depict this, such as those from Turkey, Pakistan, Tunisia, Algeria, Malaysia, Libya, Mauritania, Azerbaijan, Turkmenistan, Uzbekistan, Comoros, Singapore, the Turkish Republic of Northern Cyprus, and Western Sahara.

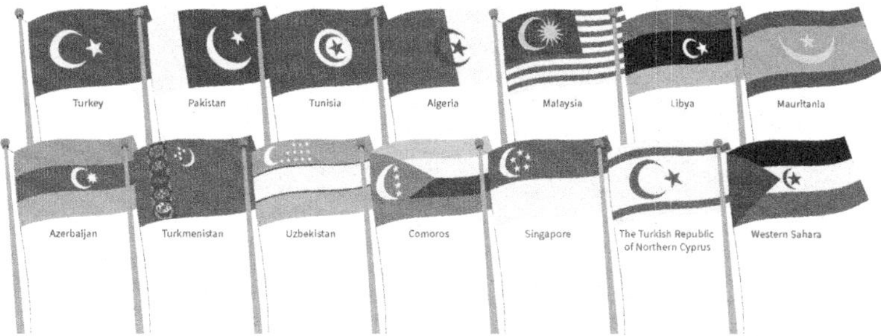

Many countries' flags show a star that is seemingly visible through a part of the Moon that is supposed to be solid. Does this indicate that cultures around the world have, at one point, acknowledged that the Moon isn't truly solid?

Some even wonder if the Moon is physically located where we see it or whether it's a projection from some other location (and the same could be said about any light in the sky). Furthermore, ancient cultures spoke of a time in which the Moon was not in the sky at all.[66]

Many possibilities about the Moon and other celestial bodies come into view once the mind is open to possibilities beyond the standard Globe model. And when the inquiry is taken a step further by investigating claims made by space agencies, questions about the cosmos solidify even more. That is where the exploration continues in the next chapter.

CHAPTER 6

SPACE AGENCIES

Igor Volk was a Russian astronaut who received a medal for being a "hero of the Soviet Union." In a 2014 television interview in Bulgaria, three years before he died, the interviewer said to Volk: "You are a hero of the Soviet Union, and we can trust you." Volk replied: **"Look, the code of honor prevents the cosmonauts from talking about what happens in space....We haven't been to space. If someone claims that we have been, it is not true. It is not the truth."**[1] [emphasis added]

Roughly nine years after that interview, the man who interviewed Volk, Stoycho Kerev, was interviewed himself. Kerev was asked in this 2023 interview whether astronauts went "to the moon." Kerev responded: "They did...but only in their wildest dreams!" The interviewer replied, "I watched a few tens of documentaries, and now I am absolutely convinced that nobody has been there, and the Americans have made the perfect hoax. I am used to making documentaries. And if I have to cover up inconsistencies—they come up all the time—I know that I won't be able to cover them up, especially on such a large scale, and especially for such a long period of time. 50 years have passed, right?"

Kerev then said: "One of the guests on my show was one of the greatest cosmonauts of the Soviet Union, Igor Volk....He [was] in Bulgaria...and during that time I was with him....This man has tested every kind of airplane that had ever been manufactured in the world. In a private conversation, this man told me we haven't been to space." The interviewer asked, "But he didn't tell you directly that [US astronaut Neil] Armstrong and company didn't go there?" Kerev responded, "Look, do you really need him to say that directly, considering that he is giving you the key?"[2]

Igor Volk's statements, and the apparent validation years later by Kerev, relate to a more general concept that is discussed in this chapter: the validity of cosmological evidence from space agencies (or the lack thereof). This topic is simply unavoidable when trying to explore the nature of the cosmos and our place in it. Government-sponsored agencies have had an effective monopoly on observations of "what's high above us," and what they present to the public is often regarded as convincing evidence for the nature of the cosmos. For instance, "What about NASA?" is a common rebuttal to Globe Skeptics' arguments. Often that reflexive question arises without having looked at any of the other matters discussed so far.

Certainly things are shot up into the sky, and we observe things coming back down. But do we know precisely what they are doing "up there" in between?

The problem, from a scientific standpoint, is that believing space agencies' "evidence" entails trusting a third party, wherein there is a limited ability to independently replicate the findings. And for such important topics, *lots* of independent replication from a wide range of parties would be desirable. Recall Steven A. Young, PhD's statement: "All NASA imagery is unfalsifiable and unverifiable."[3]

But there are even more problems from a scientific perspective. Given that the theories of gravity and general relativity appear to be inaccurate—as discussed in previous chapters—the notion of space exploration as it's commonly conceived needs to be challenged. Stated another way, the entire narrative of space exploration is predicated upon seemingly broken physics...which means the narrative itself may be broken too. As a

reminder, astrophysicist Pavel Kroupa and his colleague Moritz Haslbauer summarized the bleak situation in a 2023 article in *iai News*: "Our model of the universe has been falsified: The cosmological standard model is wrong."[4] This logically implies that at least some of the messages from space agencies, which likewise operate from the standard model, also must be wrong.

Therefore, it's necessary—in fact, it's *essential*—to examine with a critical eye what we've been told by space agencies.

Prior to the 1950s, this wouldn't have been an issue in cosmological debates because there was no such space exploration. But now it's vital. Space agencies drive much of the public's views on the cosmos and Earth's place in it.

Bearing all of this in mind, the remainder of this chapter presents a sampling of information about space exploration. The subject is simply too big to try to explain every single observation that has ever come from a space agency, but hopefully the examples present a framework that can be applied to exploring instances not discussed here. Ultimately, without having inside knowledge, we all have to make inferences and educated guesses about the broader implications. You, the reader, can make your own decision about what constitutes compelling evidence (or not).

Finally, it's important to keep in mind that we're considering organizations with immense funding. NASA, for example, has an annual budget of $25 billion (2025).[5] That's $68 million per day. One might wonder: *What can be pulled off with that level of funding?*

Trusting the Authorities

There seems to be widespread reverence for space agencies that curiously transcends other ideologies. For instance, many Americans are skeptical of the Chinese and Russian governments, but do they share that same skepticism when Chinese and Russian space agencies make announcements about their space explorations? In other contexts, these countries are consistently accused of propagandizing their citizens, but is this critique so regularly applied in the context of space?

Moreover, many people condemn Naziism, but do they condemn NASA, which was originally led by former Nazi SS officer Wernher von Braun? Why are some people quick to call political opponents *figurative* "Nazis" while in the same breath wanting to defend the credibility of an organization that was originally led by someone who was *literally* part of the Nazi party? And why are some quick to question government-sponsored health organizations like the Centers for Disease Control and Prevention (CDC), while at the same time trusting NASA? Finally, does it make sense that during the height of the Cold War in 1967, the United States signed a Space Treaty that was also signed by its alleged nemesis, the Soviet Union?[6]

An intellectually open and honest approach requires questioning everything, regardless of one's preconceptions. And when it comes to government-sponsored agencies, history shows that there is reason to ask questions. Perhaps there is a subconscious, if not explicit, belief that "they wouldn't lie" or "they couldn't pull off a deception of that kind" or "if that were true, I would have seen it on the news." The reality is that history is filled with deception from power structures. Consider a number of examples of the US government's past activities that demonstrate a lack of transparency around major events:

- *Pearl Harbor* (1941): A trove of previously confidential documents, uncovered through Freedom of Information requests, suggest that the US had knowledge of Japan's attack on Pearl Harbor before it happened. In spite of the casualties that resulted, there was apparently a strategic benefit in allowing Japan to commit the first act (as analyzed in historian Robert Stinnett's *Day of Deceit* [2000]).[7]
- *Gulf of Tonkin* (1964): The US escalated its involvement in the Vietnam War because of an alleged attack in the Gulf of Tonkin by North Vietnamese forces. However, in 2005 and 2006, previously concealed documents were released through the Freedom of Information Act. They suggest that this attack did not occur (though there

was, in fact, a separate attack two days earlier). The US expanded its presence in the conflict—using the attack that never happened as its justification. An essay on the US Naval Institute's website includes a quotation from a commander who reported: "I had the best seat in the house to watch that event, and our destroyers were just shooting at phantom targets—there were no PT [patrol torpedo] boats there...there was nothing there but black water and American firepower."[8] Nearly 60,000 Americans are estimated to have died in the war—a war that started because of an attack that didn't happen.

- *Operation Northwoods* (1962): A 1962 Joint Chiefs of Staff document known as Operation Northwoods (related to Operation Mongoose) was declassified in 2001. The document lays out a detailed plan to make Cuba look like an aggressor against the US. Although it was not enacted, the document reveals a deceptive mindset. For example, among other things, one plan was to "blow up a US ship," blame it on the Cubans, and "conduct funerals for mock-victims."[9]

- *JFK Assassination* (1964): The CIA enacted a campaign to brand as "conspiracy theorists" anyone who questioned the official narrative around the assassination of former US president John F. Kennedy. However, now, the "summary of findings" at the official National Archives, JFK Assassination Records website states: "The committee believes, on the basis of the evidence available to it, that President John F. Kennedy was probably assassinated as a result of a conspiracy."[10]

- *MK-ULTRA* (1950s–1970s): MK-ULTRA refers to mind-control experimentation on people, often without their consent. Many techniques have been reported, ranging from hypnosis to psychedelics to various forms of torture. Such techniques can "program" a person to conduct the government's desired activities and even create multiple personalities whereby the person isn't even aware of the alternate personalities. MK-ULTRA has gone under many names throughout its history but is widely accepted as a real program because of declassified documentation. Three particularly revealing books on this topic include: *Project MK-ULTRA and Mind Control Technology: A Compilation of Patents and Reports* edited by Axel Balthazar (2017); *Poisoner in Chief: Sidney Gottlieb and the CIA Search for Mind Control* (2019) by Stephen Kinzer; and *CHAOS: Charles Manson, the CIA, and the Secret History of the Sixties* (2019) by Tom O'Neill and Dan Piepenbring.[11]

❍ *9/11* (2001): The Twin Towers of the World Trade Center infamously collapsed in New York City on September 11, 2001. However, what's less often discussed is a third tower that fell about seven hours later. It was not hit by an airplane. The third building was known as World Trade Center building 7 (WTC 7), and it was forty-seven stories tall. It fell virtually straight down. The official story was that it fell because of a fire caused by debris from one of the other towers. Yet the University of Alaska, Fairbanks, conducted a four-year engineering analysis, released in March of 2020, and concluded otherwise. The document is more than 100 pages long. It states:

> The principal conclusion of our study is that fire did not cause the collapse of WTC 7 on 9/11, contrary to the conclusions of NIST [National Institute of Standards and Technology] and private engineering firms that studied the collapse. This conclusion is based upon a number of findings from our different analyses. Together, they show that fires could not have caused weakening or displacement of structural members capable of initiating any of the hypothetical local failures alleged to have triggered the total collapse of the building, nor could any local failures, even if they had occurred, have

> triggered a sequence of failures that would have resulted in the observed total collapse.[12]

The official "9/11 Commission Report" released by the US government doesn't even mention WTC 7. This is one of many anomalies from 9/11 that raise questions about the official government narrative. The lack of transparency has led to significant speculation of deception.

- *US Invasion of Iraq* (2003): The US claimed to have started a war with Iraq in large part because Iraq had "weapons of mass destruction" (WMDs). In 2005, after the invasion resulted in thousands and thousands of deaths, the US reported that no WMDs were found.[13]

The point is that it's natural to ask questions. If governments are willing to deceive in some instances, perhaps they're deceiving in other instances that haven't been acknowledged yet.

NASA's History

Although many other countries have space programs, the US's program is the most prominent. It's also the only one claiming to have sent people to the Moon. Therefore, NASA will be the focus of the space-agency discussion in this chapter. The themes and methods of critical thinking can be applied to any other such organization in other countries.

NASA's history is littered with occult influences. The organization was founded in 1958 and preceded by the NACA (National Advisory Committee of Aeronautics) in 1915.[14] In the early days, rocket scientist Jack Parsons was a leader of the Jet Propulsion Laboratory, and he was a follower of Aleister Crowley's mystical teachings. As concisely summarized by the Science History Institute in a 2024 podcast episode titled "The Sex-Cult 'Antichrist' Who Rocketed Us to Space":

> In 1939, Parsons and his wife Helen attended a so-called gnostic mass. It's a rite in a religion called Thelema.

> Aleister Crowley invented Thelema in the early 1900s. It drew on a mélange of different practices, from ancient Egypt to pagan rites to Free Masonry.
>
> Two central aspects of Thelema are magick [*sic*] and sex rituals. In Thelema, magick doesn't mean stage illusions. It's more like paranormal experiences: interacting with spirits, talking with the dead.
>
> Meanwhile, sex rituals are exactly what they sound like. In Crowley's religion, sex was considered a sacrament. And during the gnostic mass, worshippers ate so-called Cakes of Light. They're eucharists made of flour, honey, oil, ash, and often bodily fluids like semen or menstrual blood.[15]

Parsons also became close friends with L. Ron Hubbard, the founder of the Church of Scientology. As reported by Hugh Urban, PhD, a professor of religious studies at Ohio State University, Hubbard was even a "Scribe" in Parsons's esoteric sex ritual in which he chose a female partner to "become the vessel for the 'magickal child' or 'moonchild,' a supernatural offspring that would be the embodiment of the ultimate power: indeed, this child would be 'mightier than all the kings of the Earth.'"[16]

NASA also had a Nazi influence. As mentioned earlier, former Nazi SS member Wernher von Braun led NASA in its early days after being brought to the United States in Operation Paperclip, which involved bringing Nazi scientists to America, starting in 1945. The Nazis similarly had an interest in the occult, practicing rituals of various kinds[17] (some of which were later emulated by the openly practicing satanist in the US military, Lieutenant Colonel Michael Aquino, who even traveled to Heinrich Himmler's ritual chamber[18]). Whether von Braun was actively bringing such practices to NASA is not evident, but the occult connection is worth noting. More broadly, as summarized by PBS's program titled *Tainted Legacy*: "Many Americans who idolized von Braun in the 1960s knew little of his Nazi past."[19]

Along these lines, in President Eisenhower's farewell speech in 1961, he warned: "We must be alert to the…danger that public policy could itself become the captive of a scientific technological elite." He was asked by Herbert York, then-head of the Advanced Research Projects Agency, which preceded DARPA, whether he had "any particular people in mind." Eisenhower replied: "Wernher von Braun and Edward Teller [father of the hydrogen bomb]."[20]

There's a noteworthy NASA–Nazi Antarctica connection as well. Shortly after NASA's founding in 1958, the Antarctic Treaty was signed in 1959. This has functionally restricted public access to Antarctica. Interestingly, both the Nazis and the United States had an interest in Antarctica. For instance, the History Channel acknowledges Nazi explorations of Antarctica in 1938, in which they named a mountainous region called "Neu-Schwabenland."[21] And, as discussed earlier, the US had explored Antarctica via Operations Highjump (1947–1948) and Deep Freeze (1955–1956).

From a Globe Skeptic perspective, a narrative could be constructed: *They knew that Antarctica is the ice-wall perimeter of the Flat Earth, and potentially there are valuable resources in those lands, so the formation of NASA and the Antarctic Treaty were mechanisms to misdirect the public with space travel to try to keep the public ignorant of what's in the lands beyond.* Of course, there are many other possible narratives that could be constructed too. There could simply be hidden land and resources on the continent of Antarctica within a conventional Globe model.

It's also worth considering that the Central Intelligence Agency (CIA) was founded in 1947, and the National Security Agency (NSA) was founded in 1952. Many potentially secretive activities were occurring around the time of NASA's founding.[22]

Furthermore, in 1962, several years after the formation of NASA, the US launched Operations Fishbowl and Dominic. As a reminder, these involved shooting "nuclear weapons" in the sky, which Globe Skeptics speculate involved testing a "firmament" enclosure above. Wernher von Braun's grave even includes "PSALMS 19:1" under his name—and as PBS's Tainted History website explains: "Von Braun's modest grave marker refers to one of his favorite biblical passages: 'The heavens declare the glory of God; and the firmaments showeth His handiwork.'"[23] Some might argue that this refers simply to the sky, whereas Globe Skeptics wonder if this refers to the literal enclosure of the flat plane of Earth. The implication here would be that von Brahn knew that the Globe model is a lie, and NASA has been part of perpetuating that lie.

The Tainted History website also summarizes von Braun's interest in spreading the message about "space" to the public, including children: "In the 1950s, von Braun worked with Walt Disney's studios as a technical director on television programs about space exploration."[24] From a Globe Skeptic lens, such a partnership with Disney could be viewed as a means of propagandizing new generations about a spherical Earth that sits within a vast vacuum. Von Braun was known to have had a close relationship with Walt Disney himself. Along these lines, Disney's theme park also has a famous ride: *Space Mountain*.

Leading Up to the "First Man on the Moon" (1969)

The United States is the only country to have "sent men to the Moon"—twenty-four in total, including those who simply "orbited the Moon." Twelve men were actually "on the Moon." All of this occurred between 1968 and 1972. It has not happened since. No other country—not even technically advanced ones—has even sent one human into the Moon's (alleged) orbit.

The journey began publicly for the United States in 1961, several years after NASA's inception. President John F. Kennedy announced that America wanted to land a man on the Moon before the end of the decade. However, in a recording released decades later, Kennedy clearly expressed skepticism when speaking with NASA administrator James Webb. Kennedy said: "If I get reelected, we're not going to the Moon...are we?" Webb replied, "No, you're not going." Later in the conversation, Kennedy commented: "Putting a man on the moon really is a stunt, and it isn't worth that many billions."[25]

Throughout the 1960s, there were technical and practical challenges. Some of them were mentioned in North American Aviation (NAA) inspector Ronald Baron's "The Baron Report (1965–1966)" on NASA's Apollo program. He wrote: "I have been with NAA for the past sixteen months. During that time, I took the time to make notes on daily happenings. There were difficulties with people, parts, equipment, and procedures, not to mention poor safety practices and the accidents they caused."

He listed many problems, one of which seemed to sum up his views: "There is not one procedure that I can remember that was completed without a deviation, either written or oral." He compiled another list, prefaced by these statements: "The following is a list of policies that NAA should follow to make themselves the 'professional' people they should be in the first place. I am afraid the public had the wrong image in their minds when they think of project Apollo. They probably believe that everyone knows exactly what they are doing at all times. They probably also believe that the work out here at the launch complexes is done on a routine manner. They are wrong." Ultimately, he warned: "We should not compromise the safety of the astronauts just for the benefit of a schedule."[26]

James Webb expressed doubts in 1967 about the possibility of meeting the "end of the decade" deadline to reach the Moon.[27] The next year he quit. So did NASA's deputy administrator, Robert Seamans, and astronaut Wally Schirra.

Astronaut Gus Grissom was vocal in his concerns about the Apollo program. Reportedly, he even hung a lemon on the spacecraft as a symbol of his discontent. Grissom died shortly thereafter during an Apollo 1 simulation in 1967, alongside astronauts Roger Chaffee and Ed White. Space.com summarizes the tragedy:

> The crew...did a launch test on Jan. 27, which was plagued by problems such as odors in the breathing oxygen and communications problems. Grissom complained, "How are we going to get to the moon if we can't talk between two or three buildings?" At 6:31 p.m.,

> the astronauts reported a fire in the spacecraft. As the astronauts tried to open the door—a newer design that required a complicated procedure to open—officials raced over to try to get it open on their end. By the time the door was forced open, minutes had passed and it was too late for the astronauts.[28]

Many people, including Grissom's wife, have speculated that the fire was not accidental but instead that it was a murder in response to Grissom's lack of cooperation.[29] In fact, Moon-landing researcher Bart Sibrel interviewed Grissom's wife years after he died. She said that her husband told her days before his death that NASA astronauts were at least ten years away from landing on the Moon. He died two years before the planned Moon landing mission of 1969. Grissom was also apparently in the process of writing a report that was critical of the Apollo program, and CIA agents entered his home minutes after his death and confiscated the report.[30]

Yet in spite of this checkered history, the Apollo 11 mission of 1969 is said to have landed on the Moon. The mission included the now-famous astronauts Neil Armstrong, Buzz Aldrin, and Michael Collins.

Many citizens are skeptical about the Moon landing, however. A 2019 poll indicated that roughly 10 percent of Americans don't believe it happened,[31] and a 2022 poll found that 25 percent of Europeans feel the same way.[32]

The skeptics' arguments are summarized in the comprehensive 2017 documentary *American Moon* and were echoed for decades before that. For instance, Bill Kaysing—a former employee at Rocketdyne, a company that built engines for the rockets used in the Apollo program—publicly stated that he thought the Moon landing was faked because "it was technically impossible."[33] Kaysing also wrote a book titled *We Never Went to the Moon*, published in 1974. He argued that the footage on the Moon had been filmed in a studio. Ralph Reneé wrote a similar critique in his 1994 book, *NASA Mooned America*. Since then, a plethora of material on the subject has been produced, including Bart Sibrel's popular documentaries, *A Funny Thing Happened on the Way to the Moon* (2001) and *Astronauts Gone Wild* (2004).

The Van Allen Radiation Belts

A leading reason for ongoing skepticism about the Moon landings has to do with the Van Allen Radiation Belts. These are alleged to be highly radioactive areas located between roughly 1,000 and 25,000 miles above Earth's surface.[34] They were discovered in 1958 by physicist James van Allen, who wrote in a 1959 *Scientific American* article: "The discovery is of course troubling to astronauts; somehow the human body will have to be shielded from this radiation even on a rapid transit through the region."[35]

In 1961, van Allen published another article titled "The Danger Zone: Earth Is Wrapped in Deadly Belts of Radiation," in which he remarked: "Owing to the great penetrability of the high-energy protons therein, **effective shielding is quite beyond engineering feasibility in the near future. . . . All manned spaceflight attempts must steer clear of these two belts of radiation until adequate means of safeguarding the astronauts have been developed.**"[36] (Note: To Globe Skeptics, the Van Allen Radiation Belt might be code for "the firmament.") [emphasis added]

In 1968, three astronauts on Apollo 8 apparently went through the Van Allen Radiation Belts but did not land on the Moon. However, aerospace engineer Bill Wood reported that the walls of the craft were made "as thin and as light as possible" and didn't appear to have any special radiation protection.[37] Film of the astronauts on the craft, allegedly halfway to the Moon, shows one of the astronauts displaying a toothbrush that was floating in space—with no apparent concern about the Van Allen Radiation Belts.[38] In NASA's "Apollo 8 Mission Report," which is more than 250 pages long, the term *Van Allen Radiation* does not appear once.[39] However, the "Apollo 14 Mission Report" *does* clearly state that "the spacecraft traveled through the heart of the trapped radiation belts."[40] This inconsistency is striking.

Sibrel asked Apollo 12 astronaut Alan Bean if there were any ill effects from the Van Allen Radiation Belts during his travels. Bean responded, “No. Now I’m not sure we went far enough out to encounter the Van Allen Radiation Belt. Maybe we did.” When Sibrel informed Bean that the Van Allen Radiation Belts are located between Earth and the Moon, where Bean allegedly went, Bean replied: “Then we went right out through ’em.”[41]

In 2014, NASA engineer Kelly Smith gave further reason to question the Apollo Moon-mission narrative:

> I work on navigation and guidance for [NASA’s Orion project]. We are headed 3,600 miles above Earth, fifteen times higher from the planet than the International Space Station. As we get farther away from Earth, we’ll pass through the Van Allen Belts, an area of dangerous radiation. Radiation like this could harm the guidance systems, onboard computers, or other electronics on Orion. Naturally, we have to pass through this danger zone twice: once up and once back. But Orion has protection. Shielding will be put to the test as the vehicle cuts through the waves of radiation. Sensors aboard will record radiation levels for scientists to study. **We must solve these challenges before we send people through this region of space.**[42] [emphasis added]

This is curious, given that NASA claims to have sent twenty-four astronauts through the Van Allen Radiation Belts in the Apollo missions—twice on each mission, once on the way to the Moon, and once on the way back to Earth. And this occurred decades before Kelly Smith’s 2014 video—meaning that those earlier missions presumably used less sophisticated technology than what NASA currently possesses. And yet Smith says the challenges need to be solved *before* sending astronauts through this region of space?

Terry Virts, a commander on the International Space Station, similarly remarked in 2015: “Right now, we only can fly in [low] Earth orbit. That’s the farthest that we can go.”[43] In other words, where the astronauts allegedly fly right now keeps them out of the Van Allen Radiation Belts. But decades earlier, this apparently wasn't a problem during the Moon missions? Virts adds, "The Moon, Mars, asteroids, there's a lot of destinations that we could go to, **and we're building these building block components in order to allow us to do that eventually**."[44] He uses the word *eventually*...even though humans were supposedly sent to the Moon decades earlier? [emphasis added]

Contemporary NASA astronaut Don Pettit similarly remarked: "I'd go to the Moon in a nanosecond. The problem is we don't have the technology to do that anymore. We used to. We destroyed that technology, and it's a painful process to build it back again."[45] [emphasis added]

Apollo 11 Footage and Testimony

Bart Sibrel started working on the documentary *A Funny Thing Happened on the Way to the Moon* in the mid-1990s. In his book *Moon Man* (2021), he gives context for the financial backing he received:

> I met a highly successful millionaire who had made his fortune in the technology sector and who was a board member of an aerospace company building rockets for NASA....He knew, from an engineering standpoint, that the massive Saturn V rocket (the one used in the supposed attempts to go to the Moon) did not have enough fuel and power to leave Earth orbit, and the Apollo onboard guidance computer (AGC) used to interface with the ground-based mainframes...did not have the capability to convert the data for the thousands of miles per hour trajectories in real time. It is worth recalling that the AGC had just 4KB of memory, and the Lunar Module...did not have enough battery power to run the air conditioning system for three days against the lunar outside temperatures of some 250° Fahrenheit....Neither did it have any ability to shield against solar radiation, let alone an unpredictable Solar

> Particle Event. This astute millionaire knew more than I did that the alleged Moon missions were really a clever and outrageous deception. When I told him of my "one-in-four chance" estimation of the probability of Moon landing fakery, he said, "No Bart. A 100% chance."
>
> It was this wealthy person who largely financed my documentary and became its executive producer. Because of his high-ranking status with a NASA contractor and within the technology sector, he wishes to continue to stay anonymous.[46]

In 1999, while working on the film, Sibrel was coordinating with NASA directly to obtain footage and photographs. He focused his requests on Apollo 11, the first allegedly successful Moon landing. His thinking was that if the Moon landings had in fact been faked, NASA was most likely to have made errors in its first attempt.[47] He comments on what happened when he reviewed the material: "To my incredible surprise, there were actually very few pictures of the two acclaimed astronauts allegedly standing on the Moon...only about two dozen....In fact, I could not find a *single* still photograph of the most famous man in the world at the time, Neil Armstrong, claiming to be standing on the lunar surface. When I went in person to the NASA archives in search of such an image, to all of the librarians' extraordinary astonishment, there were no still pictures of Neil Armstrong standing on the surface of the Moon."[48]

Perplexed by this oddity, Sibrel began sorting through tape reels sent to him by NASA. He played one at the very bottom of the pile, and on the screen there appeared the following message (of which Sibrel provides a screenshot in his book): "This film of the Apollo 11 Mission was produced as a report film by THE MANNED SPACECRAFT CENTER and is not for general public distribution."[49] **This turned out to be unedited footage of the Apollo 11 mission, revealing apparent NASA deception that Sibrel included in his 2001 documentary.** Whether NASA sent this video as an innocent clerical error, or if someone within NASA wanted to expose fraud, we may never know.

Sibrel summarizes the footage as follows (although its essence is best absorbed by watching it in the documentary itself):

> As the astronauts were really in low-Earth orbit while they falsely claimed to be "halfway to the Moon," the window they claimed to be filming through was actually filled with the bright light of earthshine reflecting off the Earth just beneath them. **The astronauts placed a transparent color photograph (transparency) of the circular Earth, some twelve inches [or] so in width, in front of the circular window.** This transparent photo was perfectly backlit by reflecting earthshine, like a color x-ray on top of a projecting light box. **The Apollo 11 crew then very carefully inserted a black crescent-shaped piece of material on top of their one-foot transparency of the Earth to make it look like it was the dividing "terminator line" between night and day.** With the camera at the back of the spacecraft filming this, and with the interior of the spacecraft completely dark, it looked as if the Earth was floating in space at some great distance away, when it was *really* just a transparency of circular Earth in front of the brightly backlit circular window from Earth orbit! Ingenious![50] [emphasis added]

One might argue that NASA's apparent deception here, by itself, is enough to put into question all subsequent claims from the organization. If it was willing to fake such a pivotal photograph, how can any of its other images and information be trusted without significant, independent validation?

Understanding these implications, Sibrel was stunned when he saw the footage. He thought that people needed to know about it, so he planned a trip to bring it to CNN. In his book, he tells the dramatic story of getting intercepted by the CIA on his way to CNN and then being drugged.[51] After an intense encounter and eventual escape, he submitted a urine sample to a laboratory to try to uncover what he was drugged with (he suspected it was a truth-serum agent). But when he called to find out the lab results, he was informed that there was an overnight break-in, and the only thing that had been taken was his urine specimen.[52] Sibrel waited until 2021 to reveal this event, which occurred in 1999. By the time of his book launch, the footage he received from NASA had been publicly available in his documentary for about two decades already.

Shortly after his run-in with the CIA, Sibrel was intercepted by another group of CIA agents. But this time, they were apparently the "good guy" faction of the CIA. Sibrel recounts that the men from the CIA spoke to him about an internal, unpublicized war within the government itself. Some of them "know about the Apollo deception [and] find it incredibly disgusting."[53] Apparently they were monitoring Sibrel (and, reading between the lines, they were protecting him). Sibrel writes that he was able to disclose what happened in the book because the period of his confidentiality agreement with the good-guy CIA officers had ended.[54]

Sibrel also reveals testimony in his book suggesting that the Apollo 11 Moon landing was filmed at Cannon Air Force Base, near Clovis, New Mexico. The testimony comes from the son of a security officer who claimed to have been on-site at the filming location. The man's father confessed on his deathbed because he had been sworn to secrecy. However, as the man notes, "The original recording that my father made on his deathbed was destroyed by a fire." Now on his own deathbed, he decided to share his father's confession in a video recording. It is available at https://www.sibrel.com/moon-man-book-links, link #16, titled "Deathbed Confession." Among other details, the man said that his father recalled seeing the Moon landing on TV and recognized that it "was exactly what they recorded at that hangar."[55] Unfortunately, these sorts of anecdotes are difficult to verify. But the notion of a filmed Moon landing aligns with those who believe that filmmaker Stanley Kubrick produced it, alleging that his 1968 film *2001: A Space Odyssey* was evidence of his abilities.

Along similar lines, Sibrel mentions two other noteworthy anecdotes in his book. The first relates to his documentary *Astronauts Gone Wild* (2004), in which he filmed his sometimes-contentious conversations with Apollo astronauts. He asked them to swear on the Bible that they went to the Moon. For instance, when he met Neil Armstrong, he offered him $5,000 in cash that could be given to the charity of Armstrong's choice if he would simply swear on the Bible. Armstrong refused.

But Sibrel claims that he spoke with Armstong on a separate occasion in which Sibrel told Armstong "for his soul's sake" that he should come clean. Sibrel writes: "I cannot tell you exactly what his reply was for reasons of confidentiality, yet he made it clear to me that my assessment of the [faked] Moon landings is correct… and that he had his own plan for eventual disclosure. Armstrong is said to have died shortly thereafter."[56]

Armstrong has been widely critiqued by Moon-landing skeptics for his unwillingness to conduct interviews after being a "national hero" as "the first man on the Moon." His almost-depressed demeanor in the post-Moon-landing press conference has also been a point of contention.

Armstrong's post-mission demeanor raises a number of important, but speculative, ideas about what might have happened to the astronauts. Space psychologist Iya Whiteley, PhD, asks in a 2024 lecture: "What is interesting to me—we had over a dozen astronauts that went to the Moon, but have we actually heard genuine experience of what it is like looking back apart from the photograph? Have they had the language to communicate the transformation that happened to them? In fact, a lot of them have been ostracized from family, society, and they had to resort to drinking and being isolated—because they could not communicate their experience."[57]

One might wonder if some of the astronauts behaved this way because they were coerced, blackmailed, threatened, and/or bribed into engaging in deception that inspired guilt. Furthermore, given the government's known involvement in mind control, one cannot know what astronauts are subjected to, whether memories are implanted and/or suppressed, or whether they are tortured. Mind-control activities are inherently secretive, so we don't know precisely what goes on. But what has been revealed is that NASA funded the US government's psychic spying program in the 1970s. So they clearly had some interest in the human mind.[58]

From this lens, Whiteley's statements make one wonder what really has happened to these astronauts. She further remarks: "They are heroes, but nobody prepared them to be back from the Moon. They were put into a bunker, essentially. They couldn't touch their family, they couldn't hug them. Their family was afraid to hug them because they could have been harboring disease or bacteria or foreign life, so coming back to that reality and kind of being literally behind the glass—that has affected them. Some of them lost their family."[59] One might wonder: *Was the "bunker" purely about health, or was something more nefarious at play related to mind-control or other manipulative means?* Without inside knowledge, we're left to speculate.

Sibrel's second anecdote in this vein is a story that he tells of Bill Kaysing. Recall that Kaysing was one of the early Moon-landing skeptics who wrote a book published in 1974 and had worked at a company that built engines for rockets used by NASA. Kaysing, while still living, had told Sibrel that he scheduled a conversation with Apollo 15 astronaut James Irwin in 1991. Sibrel claims that on the day Kaysing and Irwin were supposed to have "a very important conversation about the Moon missions," Irwin had a fatal heart attack. Irwin had become a born-again Christian after the Apollo 15 mission, and Sibrel seems to imply that his religious change of heart made him want to come clean.[60]

Moon-Mission Inconsistencies and Anomalies

Researchers over the years have pointed out other "issues" with the Moon missions. The intent here is not to relitigate the case for each one, as inevitably arguments have been made on both sides. Instead, the purpose is to share some of the most common categories of arguments that have been used to challenge NASA's narrative; and you, the reader, can decide what to think for yourself.

First, NASA has admitted the following: **"Aside from a few canisters of Apollo 9 telemetry tapes still stored at the WNRC [Washington National Records Center], the Apollo-era telemetry tapes no longer exist—anywhere."**[61] Telemetry tapes refer to the data sent from the missions themselves. This quotation comes from a publicly available 2010 NASA document titled "The Apollo 11 Telemetry Data Recordings: A Final Report." *Wow.*

But beyond this glaring lack of transparency, there are many anomalies noted among the things that can be observed. For instance, some of the communications between astronauts allegedly on the Moon, and NASA officials in Houston, were too fast. In other words, based on the believed distance between the Moon and Houston, the communications should have had a delay, and yet the audio tapes lack that necessary delay. This suggests to skeptics that they weren't actually communicating from so far way. The documentary *American Moon* demonstrates this by playing the audio recordings from the original, unedited tapes via Spacecraft Films DVDs. The documentary further shows that the Apollo Journal website has apparently re-uploaded the audio clips and inserted more time between communications, seemingly to correct their past mistake.[62]

Moreover, the astronauts themselves have given inconsistent testimony. When Neil Armstrong was asked in the press conference following the Apollo 11 mission whether the astronauts were able to see stars, he responded, "We were never able to see stars from the lunar surface or on the daylight side of the moon, by eye, without looking through the optics." Astronaut Michael Collins, who was orbiting the Moon while Armstrong and Aldrin were on the Moon's surface, then said, "I don't remember seeing any." This is significant, because Collins was supposedly orbiting and thus was not physically on the Moon. Yet this contradicts statements from members of the Gemini missions, who allegedly flew in space a few years earlier and remember long hours that they spent looking at the stars. Astronaut James Reilly similarly commented more recently that there isn't any atmospheric distortion in space, so the stars "don't twinkle" and he saw "millions" of them.[63]

In another apparent inconsistency, Apollo 12 astronaut Alan Bean said that the descent engine was completely quiet because of "the vacuum of space," whereas Apollo 17 astronaut Eugene Cernan said that it was "very loud."[64] As Sibrel also notes, "Another one of Cernan's mistakes was that he claimed that Earth was continually visible just over the lunar horizon during his entire moonwalk… and that the Earth was four times bigger than a full Moon on Earth. Why is it then, that there is not even **one** photograph of this amazing sight; a picture of Cernan (or any astronaut, on *any* mission) standing on the Moon with the large Earth right over his shoulder?"[65]

Even the alleged space material brought back from the Moon is under review. In 2009, decades after the first Moon landing mission, the *Telegraph* reported: "A moon rock given to the Dutch prime minister by Apollo 11 astronauts in 1969 has turned out to be a fake." The article reads: "Curators at Amsterdam's Rijksmuseum, where the rock has attracted tens of thousands of visitors each year, discovered that the 'lunar rock,' valued at £308,000, was in fact petrified wood."[66] **So the astronauts gave the Dutch prime minster a gift from the Moon years ago, and it has now been shown to be fraudulent.**

Sibrel tells another related story: "A seventy-five year old woman was even arrested by federal agents for being in possession of an extremely tiny supposed 'moon rock' the size of a gnat, which was trapped inside of a thick paperweight (so that it could not be easily opened and examined), that had been given to her deceased Apollo engineer husband by Neil Armstrong as a souvenir."[67] **Casey Sullivan, JD, reports on Findlaw.com in 2019 that when the woman "contacted the agency for help selling the rice-sized rock, in part to cover her son's medical bills, NASA organized a sting, detaining her in a Denny's parking lot, and declining to let her use the restroom, even as she wet herself."**[68] [emphasis added]

From a technical lens, there are also questions about whether scientists were even capable of bringing people to the Moon. In fact, Wehner von Braun coauthored a 1952 book, *Conquest of Mars*, in which he referenced such challenges: "It is commonly believed that men will fly directly from the Earth to the Moon, but to do this we would require a vehicle of such gigantic proportions that it would prove an economic impossibility. Calculations have been carefully worked out on the type of vehicle we would need for the nonstop flight from the Earth to the moon and to return. The figures speak for themselves: Three rockets would be necessary, each rocket ship would be taller than the Empire State Building and weigh about ten times the tonnage of the *Queen Mary* [a retired British ocean liner]."[69]

And yet the 1969 Apollo mission's technology, which launched seventeen years after this statement, was very different. Sibrel expresses his skepticism: "In comparison to the above requirements, the rocket that the United States government *claimed* sent astronauts to the Moon...weighed just two-thousand five hundred tons versus the stipulated eight hundred thousand tons of the *Queen Mary*. Instead of reaching the required height of one thousand and fifty feet as per the Empire State Building, it was only three hundred sixty-three feet tall...and there was just one rocket, instead of the necessary three."[70] NASA defenders will likely contend that technological improvements over seventeen years accounted for these changes, whereas skeptics see the discrepancies as evidence of deception given the relatively short time frame.

Skeptics have similarly examined the images of the Apollo equipment used in the Moon landings and question whether it's believable that it had the technical capability to successfully complete the missions and survive the harshness of the trip (such as withstanding flying space debris that is assumed within the Globe model).

Likewise, there are anomalies related to images and footage from the Moon missions. Discussions on this topic can quickly turn into technical debates about optics and videography. Arguments and counterarguments seemingly flood the internet. In any event, what follows are some categories that are commonly discussed: whether the images suggest that there were artificial light sources, such as those expected on a film set; whether the Sun itself looked realistic; whether the Sun and Earth sizes in the images were appropriate; whether the American flag should have been waving on the Moon given the alleged gravitational conditions; whether the Moon's sand fell too fast relative to an astronaut's landing after a jump, given the alleged gravitational conditions; whether the footprints on the Moon make sense, given the alleged lunar conditions; whether there was a "Kubrick horizontal" line in the background of lunar images that were similar to what Stanley Kubrick used in *2001: A Space Odyssey*; whether the shaking of film equipment should have disrupted the transmission during "live" broadcasts given the sensitivity of the antenna's angle; whether the film equipment itself was capable of performing, given the radiation it should have had to endure; and many more. Summaries of the anomalies are covered concisely in Marcus Allen and Trevor Weaver's *The Apollo Moon Hoax: The Real Evidence* (2023); Randy Walsh's *The Apollo Moon Missions: Hiding a Hoax in Plain Sight, Parts I and II* (published in 2018 and 2021, respectively); the documentary *American Moon* (2017); and many other sources.

"Too Hard to Fake"

Often there's a tendency to dismiss the anomalies as inconvenient data points that can be swept under the rug. As the thinking goes, *There's no way so many people could be in on it.* The problem with this argument is that many individuals involved in the missions are contractors who are compartmentalized—they work on individual pieces without knowing the full situation. Often they're even in different geographies, building different physical structures.

But most fundamentally, the number of direct observers of space phenomena is very, very small; and that leaves open the possibility of distortion and deception. That's why many independent replications are so critical. And yet with "space," that's not easy. **As Sibrel puts it: "There were only three people who were actually supposed to be on and around the Moon at the time of each Apollo mission, with no independent press coverage."** He adds, "Just like a pyramid of power in any business, what the employee, the manager, and the regional manager know about the business's actual activities and agenda is completely different than what the

CEO at the very top knows."[71] Furthermore, those involved in important projects typically have to sign nondisclosure agreements that legally prevent them from discussing their work. [emphasis added]

Sibrel further comments that he's spoken to people at NASA who "know the Moon landings are fake." **But they cannot say it officially or publicly because they'd get fired.** In Sibrel's words, "It's a Catch-22" for these NASA employees.[72]

Also, the sophistication of NASA's simulation technology raises questions. Gene Kranz, a former NASA flight director, wrote in *Failure Is Not an Option: Mission Control from Mercury to Apollo 13 and Beyond* (2000): "**In the late 1960s our simulation technology had progressed to the point where it became virtually impossible to separate the training from the actual missions.** The simulations became full dress rehearsals for the missions down to the smallest detail. The simulation tested the crew's and controllers' responses to normal and emergency conditions.... We simulated every mission phase under a variety of normal and emergency conditions."[73] [emphasis added]

In other words, as authors Allen and Weaver ask: "Could the simulations have been realistic enough to convince the general public and the mission controllers?"[74] They also quote NASA engineer Jack Clemons, who worked on the Apollo project in the 1960s: "With those [simulators] we could simulate most of the mission so that during the actual mission you had these computers crunching along, telling us how it was going and what needed doing." And Apollo 15 astronaut Alfred Worden said: "I must admit that, at the cost of appearing insensible, we trained so intensely and for such long periods that when the moment of the mission came, it was like carrying out another training session in the simulator."[75]

Finally, a woman who worked on Apollo 8 and 10 said, "The simulations feel as real as the mission." She was then asked whether the program has a sense of "unreality" to it, to which she replied: "I think the atmosphere you work in is a little bit unreal as far as the actual facilities are concerned. You're totally isolated. You're surrounded by machines. There are no windows in the building where you work. The lighting is always the same. You lose all track of time."[76]

Given all of these anomalies, perhaps the following remarks from former US president Bill Clinton make sense. They are from his autobiography, *My Life* (2004):

> Just a month before, Apollo 11 astronauts Buzz Aldrin and Neil Armstrong had left their colleague, Michael Collins, aboard spaceship *Columbia* and "walked on the moon," beating by five months President Kennedy's goal of putting a man on the moon before the decade was out. [An] old carpenter asked me if I really believed it happened. I said sure, I saw it on "television." He disagreed; he said that he didn't believe it for a minute, that "them television fellers" could make things look real that weren't. Back then, I thought he was a crank. During my eight years in Washington, I saw some things on TV that made me wonder if he wasn't ahead of his time.[77]

Problems with Images from Above

If NASA's history and Moon-landing anomalies aren't concerning enough, consider that NASA has not released a complete, nondoctored image of Earth since 1972. This image was from the Apollo 17 Moon mission and is allegedly the only one of the whole Earth taken by humans. But even to call it a "whole Earth" image is misleading, because it only shows one side—so it's more like a disk. In 2015, NASA said that it released its first image of the "whole Earth" in forty-three years, but even that was a "composite" of three separate images.[78]

Within the context of the Globe Skeptic discussion, this raises red flags. One would think that there should be thousands and thousands of complete pictures that demonstrate the "oblate spheroid" shape. One would expect that there would be many, many pictures that aren't composites or computer-generated imagery (CGI). And shouldn't there be pictures taken of Earth showing buildings standing right-side up on the top of the sphere, while simultaneously being held sideways on the side of the sphere, and upside-down on the bottom? And might we expect some complete videos of the full Earth from a distance, showing all of its geographic locations, *in continuous footage*, as it rotates?

These issues naturally raise questions about space-agency pictures of every other celestial body, not just Earth. Clearly, the technology exists to make something *look* real when it's not. Since the images we're shown cannot be verified or falsified by the average citizen, they're unconvincing to skeptics. The question that any intellectually honest person would ask about *any* picture coming from "space" should be: "Is it real, or has it been doctored in any way? And how can I be sure?"

Consider the 2002 rendition of the 1972 Apollo 17 image of Earth, known as "The Blue Marble." It appeared on Apple's iPhone. It was *everywhere* and branded into the public psyche a specific image of Earth from space.

The famous 2002 "Blue Marble" image of Earth, also available at https://science.nasa.gov/resource/blue-marble-2002/. It was created by NASA's Robert Simmon, who said it was "photoshopped."

Its creator was NASA lead data visualizer and information designer Robert Simmon. In a 2012 interview, he remarked: "My role is to make imagery from Earth sciences data. I turn data into pictures. I look for new, interesting events that NASA's satellites have seen or that are hidden in the latest data to find anything interesting that shows off NASA's unique capabilities. Finding things is the fun part. I rely on engineers and scientists to produce the data." He adds:

> The last time anyone took a photograph from above low Earth orbit that showed an entire hemisphere (one side of a globe) was in 1972 during Apollo 17. NASA's Earth Observing System (EOS) satellites were designed to give a check-up of Earth's health. By 2002, we finally had enough data to make a snapshot of the entire Earth. So we did. The hard part was creating a flat map of the Earth's surface with four months' of satellite data. Reto Stockli, now at the Swiss Federal Office of Meteorology and Climatology, did much of this work. **Then we wrapped the flat map around a ball. My part was integrating the surface, clouds, and oceans to match people's expectations of how Earth looks from space**. That ball became the famous Blue Marble.[79] [emphasis added]

Separately, Simmon stated in an interview that his job is "primarily taking data and making pictures out of it....To us the really cool thing was the dataset. Up until that point, there was no realistic color map of the globe anywhere." He describes how, in creating the Blue Marble, some of the images came from an instrument that measures phytoplankton. This affected the way he would design the final image of Earth. He ultimately colored Earth's water based on phytoplankton quantities: "Where [the phytoplankton count] was low, I colored it dark blue because they're low mostly in mid oceans, and then where it was a little bit higher, it was like a little bit brighter green....There's a small problem with it because there's a very slight gap in between each orbit....**It is photoshopped. But [it] has to be.**"

Then he describes "another layer to sort of simulate the atmosphere," and he mentioned the important detail of managing the "specular highlight, [which is] the reflection of sunlight off of water." He says piecing it together initially "didn't look realistic; it looks kind of flat or the clouds are sort of too see-through, so I just take 'Command Z' a lot [referring to the "undo" function on computers]." **The final version of the Blue Marble image was, in Simmon's words, "what I imagine it to be. Unfortunately, I'm not an astronaut. I've never been to space, but I've looked at these images over and over again trying to sort of get the essence of it." As the interviewer says: "There's artistry to creating the world."**[80] [emphasis added]

Lens technology also introduces problems. For instance, consider what a photography-guide website explains about the *fisheye camera lens*: "A fisheye lens is a special type of ultra-wide angle lens. **It...shows a distorted, spherical view of the world**, most evident in the curved outer corners of the photo, known as the 'fisheye effect.'...When reading about fisheye and regular wide angle lenses, you will hear a term called *barrel distortion*. This distortion causes curved lines at the edges of the photo....However, on fisheye lenses, this is their main feature. That is why they are called 'fisheye,' and the barrel distortion should not be viewed negatively."[81] [emphasis added]

These issues apply to space photography. For instance, Smithsonian's National Air and Space Museum website shows a camera and states: "This Nikon fisheye lens was used on the STS-30 and STS-31 (Hubble deployment mission) space shuttle missions in the early 1990s with a Hasselblad camera, **one of the most commonly used cameras in the space program**."[82] Pictures from the International Space Station (ISS) have used fisheye lenses, such as a 2019 picture taken by astronaut Andrew Morgan, which is labeled by NASA's Earth Observatory website as "Fisheye over Sinai."[83] *Time* magazine even wrote an article in 2016 titled "Take a Tour of the International Space Station in Dazzling 4K Detail," of which the first sentence reads: "NASA has created a new Ultra High Definition (4K) 18-minute video of the ISS, using a fisheye lens for extreme focus and depth."[84]

This technology clearly existed back in the Apollo mission era because NASA includes on its website an article with an image of Apollo 9, and the caption reads: "This fish-eye camera lens view of the interior of the Apollo Lunar Module Mission Simulator at the Kennedy Space Center is one of several selected by the Apollo 9 crew to appear in '*Apollo: Through the Eyes of Astronauts.*'"[85] [emphasis added]

Along these lines, consider a non-NASA example. In a 2012 stunt by Australian skydiver Felix Baumgartner (in partnership with Red Bull), he ascended to 127,852 feet in a helium balloon and a capsule. He jumped and broke world records. Earth can be seen in the background with significant curvature. However, a fish-eye lens was used.[86]

Ultimately, discerning truth in these situations will become more and more challenging as technology advances. **Twitter founder Jack Dorsey summarizes the problem well: "The way that images are created, deep fakes, videos—you will literally not know what is real and what is fake. It'll be almost impossible to tell."[87]**

Telescopes, Probes, and Satellites

Given the photoshopping and lensing issues discussed, it's hard to know if what we're seeing from telescopes has been doctored—including images from the Hubble and James Webb telescopes. The same questions go for probes allegedly sent into space; there are unverifiable elements that would need replication by many nongovernmental parties to validate the information we receive. Along these lines, any images from the alleged surfaces of other celestial bodies would need to be validated. The desertlike conditions of Mars, for example, could theoretically be taken on Earth locations—not to mention the possibility of photoshopping. In that vein, consider that the very first search result on NASA's Photojournal website for pictures from the 1989 Galileo probe launch to Jupiter is labeled: "Galileo Over Io **(Artist's Concept)**" (Io is one of the moons of Jupiter).[88] NASA's website includes many other images without the "artist" label, but the same questions of verification arise. [emphasis added]

Furthermore, it's possible to obtain data about celestial bodies without floating in space. The Stratospheric Observatory for Infrared Astronomy (SOFIA, 2010–2022), for instance, flew in the sky with a modified Boeing 747SP from which it made its observations. There was also a Balloon-borne Large Aperture Submillimeter Telescope (BLAST; 2005, 2006, 2010, 2018) which was lofted by high-altitude balloons and took pictures of stars and galaxies. So the technology exists to take pictures even when not in "outer space."

This discussion often leads to questions about "satellites": What are they? How do they stay in the sky?

It's important to acknowledge that NASA's first satellite was taken up on a balloon. Launched in 1960, it was called Echo 1, reportedly reaching 1,000 miles above (which is four times higher than the International Space Station's altitude). The craft itself was nicknamed a "satelloon."[89]

Balloon technology might be more prevalent and important than many of us realize. NASA even has a "Scientific Balloons" website.[90] A 2018 MIT Technology Review article titled "The US military is testing stratospheric balloons that ride the wind so they never have to come down" remarks: "It's not a new idea. Indeed, the original stratospheric balloons were flown by NASA in the 1950s, and the agency still uses them for science missions."[91]

Furthermore, a 1960s CIA document, declassified in 2004, discussed a "program goal" to develop a "balloon reconnaissance" system that could fly up to 200,000 feet.[92] A RAND Corporation technical report similarly noted that the US Army used high-altitude airships and airplanes flying above that could help with "communications."[93] Google's 2015 "Project Loon" was similarly set up to deploy balloons for the purposes of internet communications.[94]

We've been conditioned to think about the importance of communication and navigation technology "above us." But there is essential land-based technology as well. A 2015 *Newsweek* article notes that "99 percent of all transoceanic data traffic goes through undersea cables, and that includes Internet usage, phone calls and text messages."[95] Terrestrial cellular towers are abundant too. Also, while satellite phones are sometimes promoted as an effective communication means because of their broad communication coverage, they do in fact face connectivity challenges sometimes.[96] The point is that the entire matter of satellites and their role in communications might be more complex than is typically acknowledged.

In summary, Steven A. Young, PhD, expresses his skepticism of the traditional satellite narrative:

> People see lights moving in the sky and falsely believe them to be "satellites," but according to the globe model, the minimum height for a satellite to achieve orbit is about 180 miles. It is simply not possible to see a small device in the sky at night from 180+ miles away. Whether it has lights on it or not, the angular resolution of the eye is exceeded. I am however well aware that there are mysterious "UFO" moving lights in the night sky; I've seen them. I don't know what they are, but I know they are not tin cans in free fall at 180 miles away (Sky TV satellites are said to be orbiting at 26,000+ miles!).[97]

Therefore, it's possible to acknowledge that advanced technology exists in the sky, but perhaps the story is complicated. Drones, balloons, airplanes, and other technologies could be lights that we see "up there." Perhaps some of them help with communications, and others don't.

Or maybe, in some cases, satellite technology is indeed "floating" up there—but the floating is due to advanced technologies that are not known by the public. Aether-based technologies and electrogravitics are considerations among Globe Skeptics, for example. Young's reference to UFOs could theoretically encompass examples like those.

International Space Station (ISS) Video Anomalies

A final question about "space" that needs to be addressed is the ISS. For context, the ISS has been around since 1998 and is said to be roughly 260 miles above Earth. This places the ISS within the thermosphere but below the Van Allen Radiation Belts. Claims that still images or footage from the ISS prove that Earth is a sphere would need to be evaluated for the same reasons discussed earlier. Is photoshopping being used? What kind of lens are they using? Is the "curve" that they show consistent with the specific amount—and type—of spherical curvature that should be there, based on the Globe's radius value, at the ISS's altitude? Does imagery or continuous footage of the *entire* Earth exist, or are we continually shown snippets of little parts of Earth?

But even beyond all that, careful investigation of the ISS's footage reveals repeated anomalies, according to NASA skeptics. For instance, in April 2023, activist and public speaker Justin Harvey brought forth evidence of what he calls "potential fraud of an enormous scale" to a Brevard County, Florida County Commission Board meeting. He urged the panel to investigate this, since NASA has a significant presence in that county. His recorded public comments went viral because he concisely synthesized the anomalies that are causing a growing number of people to question whether the official story about the ISS is truthful.[98] (Note: Harvey's comments also went viral in May of 2024 after he presented anomalies to the Brevard County commission related to the Challenger explosion in 1986.[99])

With modern technology, it's always difficult to know whether background scenery in a video is legitimate or if it's computer generated. Even on simple Zoom (video) conference calls, this technology is employed, whereby people can look as if they're in a totally foreign setting. One can only wonder what kind of sophisticated technologies government agencies have. In fact, the CBS sitcom *The Big Bang Theory* included a 2012 episode with a fake ISS scene, in which the actor is floating around in a realistic

ISS-looking set.[100] Also, in 2018, a Tesla vehicle was allegedly sent into space and was shown floating above Earth. As Elon Musk said in a press conference, "You can tell it's real because it looks so fake."[101]

In one ISS video, a female astronaut is shown floating inside of the ISS, and says into the microphone, "I want people…to understand why the science on the ISS helps us out, here on Earth." She said "here on Earth" when allegedly in the thermosphere.[102] As Austin Whitsitt critically remarks, "I thought you were in space." This raises questions about whether she was simply in a studio simulation on Earth but accidentally "slipped" with her words. However, if one wanted to be generous, one might conclude that she included the thermosphere as part of Earth.

The next ISS anomaly is harder to explain away, though. As an astronaut is shown floating on the ISS with two other astronauts, he told a story that concluded with: **"And all of that happened in a little town called York, Maine, across the United States from where we're talking to you right now."**[103] Were the astronauts actually in, say, Houston, while pretending to be on the ISS? [emphasis added]

In another ISS video, it can be clearly seen that an astronaut is being held up by a harness as he swings around a corner in the background.[104] The implication is that the "floating" we see in videos might not always be due to "low gravity." Instead, apparent "floating" might be induced by physical harnesses that are used in a normal environment. And there are many video examples that skeptics feel are suggestive of harness use.[105] In fact, there is a paper coauthored by a NASA employee titled "Practical Applications of Cables and Ropes in the ISS Countermeasures System." Thus, it's clear that such technology is used in at least some circumstances.[106]

There are also cases in which things inexplicably fall while in an allegedly "low gravity" environment on the ISS. In one video, the astronaut explicitly addresses the issue. As he's floating, he mentions, "Sometimes you find a pocket of gravity." He keeps one hand next to a hammer that's floating in the air, and simultaneously he lets go of a playing card in his other hand. The card falls straight down while the hammer is still floating. Austin Whitsitt analyzed the tape and asks a good question: "If you found a pocket of gravity, aren't you currently moving 17,500 miles per hour—so how could you determine beforehand that you were going to actually be within that pocket of gravity?"[107] The astronaut would have needed to know that this tiny pocket was specifically located where it was, and he would have also needed to know that the floating hammer by his other hand was *not* in that pocket of gravity.

But in most video anomalies, the astronauts don't provide a "gravity pocket" caveat. And objects fall anyway, even though they "should be" floating. In one such video, the astronaut was signing papers, and he dropped the clipboard, which fell straight down.[108] When he dropped it, he looked around as if to make sure no one saw what had happened. It was a clear look of guilt. Similarly, in another video, an astronaut unscrewed something on the wall, and the screw fell to the ground.[109] There are many other similar examples as well.[110]

One such case is more subtle but very telling. An astronaut is shown floating while his microphone is floating too. On the wall, several feet away from him, is a plastic bag. In that bag an object can be seen to start slowly falling *down*. So it's falling *down* while the astronaut and microphone are *floating*. Whitsitt speculates that the astronaut might be on a parabolic flight that had unavoidable turbulence, which caused the mishap on the wall.[111] Parabolic flights can be used to bring about "zero gravity" environments.

There is also a video of an astronaut floating while brushing his teeth. The video footage is fast-forwarded so that viewers don't need to watch him brush his teeth for such a long time. However, as this is happening, there is a bag floating nearby *that doesn't change its speed*. If the whole video were fast-forwarded, then *the entire scene* should have moved at the same rate. But that's not what happened. Perhaps this suggests that part of the visual "background" was not really where the astronaut was being filmed. This has led to speculation that augmented reality technology is sometimes employed and visually deceives viewers.[112] NASA did in fact publish an article in 2021 titled "Nine Ways We Use AR [Augmented Reality] and VR [Virtual Reality] on the International Space Station."[113]

In one seemingly egregious mistake, an astronaut magically appears in an empty ISS room. The background has continuity all the while, which suggests that this was the product of video imagery. In fact, his body can be seen to fade in, which suggests that it's a visual effect. In other words, there could be a "green screen" with CGI layering technology that makes the astronaut look like he's part of the scene.[114]

There are also videos in which potential "bubbles" can be seen floating in space while the astronaut is filmed allegedly working on something in space.[115] Skeptics speculate that such instances were faked space scenes that were really filmed under water—because NASA astronauts admittedly train in the "Neutral Buoyancy Laboratory" underwater.[116] Adding fuel to the skeptics' fire, astronauts have had helmet "leaks" during "spacewalks" on the ISS. A 2022 *New York Post* article reads: "Panicked NASA cancels spacewalks after ISS astronaut's helmet 'fills with water,'"[117] and a 2022 CNN article reads: "NASA review underway after water leaks into astronaut's helmet."[118]

The video "Bloopers From Space—Top 25 All Time Favorites" on the YouTube channel Flat Earth and Coffee shows some of the examples discussed here. Beyond that, researchers have compiled an extensive body of documentation of similar cases.[119] In any video or image analysis, there's always the possibility of a technical glitch that's misinterpreted as fraud. But when there are so many anomalies, skeptics contend that it's hard to "explain away" all of them.

Any single example of fraud naturally puts everything into question. As the saying goes in US law: "Fraud vitiates everything it touches."[120] [emphasis added]

PART III
HOW DOES METAPHYSICS RELATE TO OUR PLACE IN THE COSMOS?

CHAPTER 7
THE IMPORTANCE OF CONSCIOUSNESS

So far, I've presented pieces of evidence that challenge conventional views about the physical nature of the cosmos. However, the *metaphysical* underpinning has been left out. This happens all too often in such discussions. But that's not surprising given that physics is inherently focused on the physical world.

Why is metaphysics important? Because it sets the container in which the cosmos sits. An attempt to understand our place in the cosmos is ultimately futile in the absence of a sound metaphysical foundation. One might even wonder if modern cosmology is so deeply flawed because of false metaphysical premises—in addition to the many other problems already discussed in this book.

In this chapter, I explore a metaphysical framework centered around consciousness, specifically. And then in the following chapter, I apply this understanding to a discussion of our place in the cosmos—and the potential that we exist within an ongoing, multidimensional "spiritual war."

Realism vs. Idealism

Realism is a form of metaphysics purporting something so basic that most of us probably haven't thought to acknowledge it. It says that there exists a physical world "out there" that's independent of our perceptions. Put another way, it suggests that there is a full cosmos that would exist in the absence of consciousness.

What is consciousness? It could be regarded as the part of us that *experiences*. If I say, *I have a thought* or *I am awake* or *I am eating*, the "I" in those statements is consciousness. Likewise, it could be regarded as our subjective, inner awareness—the nonphysical part of us that is required to experience life. In fact, consciousness is required for us to even be able to ask questions about the cosmos.[1]

Realism tells us that consciousness is dependent on the cosmos. If there were no cosmos, then there would be no consciousness. And if all consciousness were wiped out, then the cosmos would continue on its own.

Although realism *could* be true, it is problematic from a philosophical perspective. More specifically, it lacks "parsimony"—meaning that it violates Occam's Razor ("the simplest solution is usually the best"). **Here's why: A cosmos outside and independent of consciousness cannot be validated in the absence of consciousness. Such a cosmos cannot be experienced, by definition. That's because consciousness is *required* in order to experience anything. However, realism says that the cosmos exists even if consciousness were to be wiped out. It therefore asserts the existence of a cosmos that can never be experienced or verified by consciousness.**[2]

The feeling of being conscious, on the other hand, *is* experienced. It's what you're experiencing right now. It's *known*. Conversely, a cosmos that allegedly exists beyond experience is inherently *unknown and unknowable*. It relies on a fundamental inference that something persists in the absence of consciousness. Thus, realism requires a foundational leap of faith. The irony here is that realists often strongly advocate for "science," when their own metaphysics rests on top of faith.

Albert Einstein was a realist, but he honestly acknowledged the implications. When speaking with the Nobel Prize–winning Bengali mystic Rabindranath Tagore in 1930, he said: **"I cannot prove that my conception is right, but that is my religion."**[3] [emphasis added]

An alternative framework to realism—one that is more parsimonious—*starts* with consciousness itself. It says that the full cosmos emerges *within* consciousness. This is generally known as philosophical *idealism*. It's consistent with our experience. All reality that we know exists within our consciousness.

The argument in favor of idealism is that it can explain reality at least as well as realism with fewer assumptions. In other words, it starts with the one thing we know, which is our sense of experiencing (that is, consciousness) and explains all aspects of reality as a manifestation of consciousness. This philosophy does not need to invoke a world "outside consciousness," which can never be validated or verified because it's always, by definition, "outside consciousness."

Such a philosophy could lead to *solipsism*—the notion that your individual consciousness is all that exists. Strictly speaking, this view does rely on the fewest assumptions. However, a slightly refined version expands on solipsism with a slight inference, which is that *other* people (and organisms) are conscious too. So the world isn't composed of just you, in your own world, with a bunch of zombies everywhere (as solipsism would entail). Instead, we do indeed live in a world of other conscious beings. This philosophy has been detailed by philosopher Bernardo Kastrup, PhD, in many books such as *The Idea of the World* (2018).[4]

Kastrup argues that the inference about other people's consciousness is reasonable because we already know what consciousness feels like, since we feel it. Using more philosophical lingo, we *know* the "ontological category" of consciousness. And since we know it, and we see others acting in a similar way, it's a relatively minor leap to infer that they are conscious too. Realism, on the other hand, asks us to believe in an entire reality outside consciousness that's never been experienced or verified. Thus, it asks us to conjure up an ontological category that is not, and cannot, be known—and has never been, nor ever could be, experienced. Therefore, Kastrup contends, idealism is preferred over realism.

This perspective has been gaining momentum in recent years. For instance, Federico Faggin, a key inventor of the microprocessor, endorses idealism in his book *Irreducible: Consciousness, Life, Computers, and Human Nature* (2024).[5]

Kastrup's version, known as *analytic idealism*, essentially contends that there is a universal consciousness, and each of us is a part of it. That one universal consciousness has "dissociative identity disorder," whereby each of us feels like a separate consciousness, even though we're fundamentally interconnected within a broader consciousness.

He uses the metaphor of an infinite stream of water, in which each individual is a whirlpool. Paradoxically, we are separate while still being interconnected. We exist in the same ultimate reality because we're part of the same ultimate consciousness, and we're simply experiencing different "parts of the stream."[6]

The visual shown above is an illustration of Dr. Bernardo Kastrup's metaphor. He compares consciousness to an infinite stream of water. Each of us is an individuated whirlpool within the stream, and yet at the same time we are fundamentally interconnected because we are part of the same stream. Each whirlpool is thus akin to an individual mind, which is part of the broader, overarching mind.
For more on this metaphor, see Dr. Kastrup's book titled *Why Materialism Is Baloney.*

This form of metaphysics implies that humans are vessels or vehicles that house slivers of the universal consciousness. Some might call those slivers "souls." But our ultimate identity is not the body. Our identity transcends the body.

Furthermore, the notion of "God" could be regarded as the infinite stream that underlies all of reality. It's not a *separate* being but rather the *ground* of all being. **From this lens, what we call "the cosmos" is an emergence within consciousness itself. Or, using religious language, the cosmos emerges from God.**

The aether could be regarded as a spiritual interface through which consciousness manifests into the (apparently) physical cosmos. Steven A. Young, PhD, makes a similar remark when he describes the toroidal vortex structure that engulfs the Flat Earth (discussed in chapter 4 on the topic of Aether Cosmology). He writes: "Physicists being solely focused on the physical, are always seeking a material solution to a spiritual problem. Aether provides a spiritual source from which all material manifestation results through the dynamic power of the vortex."[7]

Young also invokes the vortex within the context of the whirlpool metaphor: "Consider a water vortex, the whirlpool. Though it is separate and distinct from the rest of the water, with its own size, speed, power, etc., it is still part of the ocean. It is individual and unique, but also still connected and at one with the whole body of water. The vortex demonstrates how something can be perceptually distinct, free to move and interact but without being disconnected from the medium that manifested it."[8]

Brain Bias

In spite of the philosophical problems discussed above, realism remains compelling because of our senses. It certainly *feels* like there's a world out there that would persist if all forms of consciousness suddenly disappeared. But that view is based on the way our bodies perceive and comprehend, which has inherent limitations.

Additionally, realism likely thrives because of assumptions about the brain. We know from neuroscience that chemical and electrical activity in the brain have tight correlations with our conscious experience. But neuroscientists often jump to the conclusion that the brain is the *reason* we are conscious. Thus, they assume that the brain *creates* consciousness. However, this thinking is logically problematic because it ignores all the other possibilities that could explain neuroscientific observations. **Correlation does not necessarily imply causation.** By analogy, do we conclude that firefighters cause fires simply because these individuals are found at the scene of a fire?[9]

An alternative explanation of the brain is that it acts more like an antenna/receiver/transmitter. Or it's a processor, a filtering mechanism: it's an apparatus that allows consciousness to flow through the body. This is not to say that the brain is unimportant. Under this framework, the brain is essential to the way we have specific experiences. It mediates our experience as an interface. But it's not *everything*.

In fact, there are a number of experiences that challenge the brain-centric view of reality. Near-death experiences are instances in which a person has a "realer than real" conscious experience—sometimes reporting omnidirectional, 360-degree vision—while having a brain that's either barely functioning or fully "off" (for instance, during cardiac arrest). Sometimes getting the brain *out of the way* somehow allows *in* more consciousness. So perhaps the brain ordinarily gets in the way of higher states of consciousness—much like a blindfold.[10] **The University of Virginia's Bruce Greyson, MD, explains what this means in the context of near-death experiences: "We're left with this paradox that at a time when the brain isn't functioning, the mind is functioning better than ever."[11]** [emphasis added]

Many other phenomena challenge the brain-based view of consciousness. For example, consider a case reported by Dr. John Lorber (1915–1996) in which a student had an IQ of 126. But a CAT scan revealed that he had virtually no brain. Instead, most of his brain was filled with cerebrospinal fluid (a condition known as *hydrocephalus*).[12]

There are even cases of organ transplant recipients who take on their donor's memories and preferences. In other words, aspects of consciousness are transmitted without a brain being involved. Paul Pearsall's book *The Heart's Code* (1999) describes many such cases. One young girl received a heart from someone who had been murdered and began to have nightmares about the murder. Upon speaking with a psychiatrist, details were collected, and the heart donor's murderer was arrested—simply from the recipient's memories of the murder.[13]

Nonlocal Consciousness

There are additionally many "psychic" phenomena suggesting that consciousness is not confined to the brain—meaning that it is "nonlocal." It can exist beyond the confines of space and time. Such phenomena include remote viewing (perceiving things far away with the mind alone), telepathy (mind-to-mind communication), precognition (knowing or sensing the future before it happens), and psychokinesis (mind impacting matter without physical contact). **Repeatedly, these phenomena have been demonstrated with statistically significant results that exceed the six-sigma threshold. That is, the odds that the results were found because of "mere coincidence" were more than a billion to one.**[14] Such results have been published, for example, in Dr. Etzel Cardeña's 2018 paper in *American Psychologist*, the official peer-reviewed academic journal of the American Psychological Association.[15] This category of research is summarized well by Jessica Utts, PhD, the 2016 president of the American Statistical Association. As she wrote in 1995: "Using the standards applied to any other area of science, it is concluded that psychic functioning has been well established."[16] **Collectively, the evidence suggests that humans have innate psychic capacities, even if they are sometimes subtle and require statistics to detect them. The broader implication is that we haven't been told the truth of who and what we are.**

Various additional lines of evidence suggest that consciousness continues after the physical body does. That is, when the brain stops functioning, consciousness continues. Near-death experiences are sometimes reported with an out-of-body experience in which the person's consciousness accurately perceives something from a vantage point outside the body. Upon being resuscitated, the person gives an accurate report of what happened, indicating that the experience was not merely a hallucination caused by a dying brain. An accurate perception is, by definition, not a hallucination. Jan Holden, PhD, a professor at the University of North Texas, notes that this phenomenon seemingly "wipes out" any physical explanations of such near-death experiences.[17] Perhaps consciousness can truly exist without a brain at all.

The book *The Self Does Not Die: Verified Paranormal Phenomena from Near-Death Experiences* (originally published in 2016 and updated in 2023) includes more than 100 cases of such near-death experiences. If the conventional view of consciousness were correct, there should be *none* of these cases. Recall that a single black swan invalidates the law that "all swans are white."

Similarly, there are emerging results with studies on mediums—people alleging to be able to communicate with the deceased. The Windbridge Research Center has conducted preliminary studies using five levels of blinding, showing that some mediums are able to know things about dead people that cannot be explained by ordinary means.[18] And finally, there are more than 2,500 cases of young children who have memories of lives that are not their own. Sometimes they have birthmarks or physical defects that relate to the deaths in their "past lives." In the most compelling cases, researchers have found historical or medical records validating the children's claims. These cases have been studied most famously by the Division of Perceptual Studies at University of Virginia's medical school since the late 1960s (by Drs. Ian Stevenson and Jim Tucker).[19] In some instances, children also report "intermission memories" during a period "between lives" in which they sometimes recall choosing their parents before birth.[20]

Flipping Metaphysics on Its Head

This brief summary of evidence, which has been covered much more extensively in my previous works, points us away from the "brain bias." It challenges the idea that the brain creates consciousness. This further implies that consciousness isn't just a product of the physical cosmos.

"Black swan" consciousness anomalies likewise challenge the associated metaphysical frameworks known as *materialism* and *physicalism*. Both of these, which are closely connected to *realism*, are grounded in a fundamentally *physical* reality. All three metaphysical terms—*realism*, *materialism*, and *physicalism*—are essentially different ways of describing the paradigm under which modern science operates. Therefore, to shift our views about consciousness is massive. Arguably this is a meta-paradigm shift because we're talking about the metaphysical paradigm that underlies all other paradigms in science.

An alternative form of metaphysics that emphasizes consciousness isn't new, however. Many mystical traditions throughout history have thought this way for a long time. Also, it's been alluded to by Nobel Prize–winning quantum physicists. For example, Max Planck remarked in 1931: "I regard consciousness as fundamental. I regard matter as derivative from consciousness. We cannot get behind consciousness. Everything that we talk about, everything that we regard as existing, postulates consciousness."[21] Similarly, Nobel laureate Erwin Schrödinger said: "In truth, there is only one mind."[22] What follows is an illustration adapted from the work of Dean Radin, PhD, showing this very idea.[23]

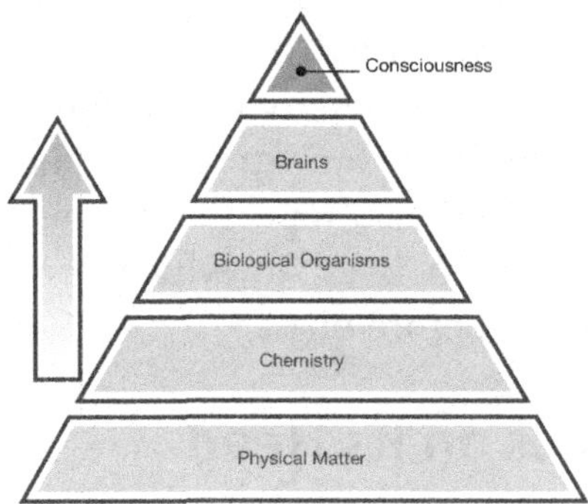

Physicalism: The interactions of units of matter (via chemistry) create biological organisms like human beings, which develop brains, out of which consciousness arises. This worldview supports atheism.

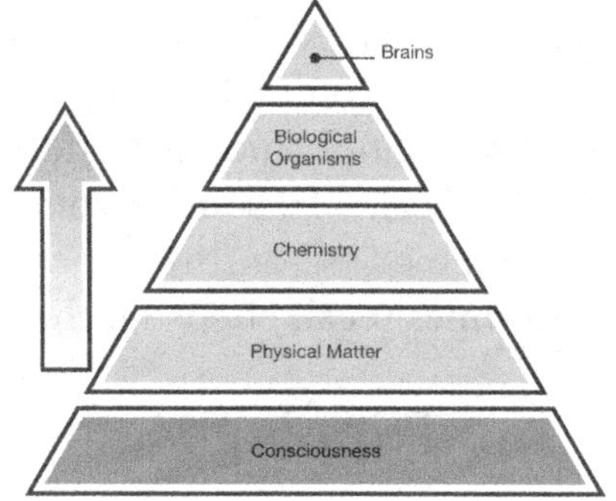

The "One Mind" view of reality: Consciousness is primary; everything we perceive in the apparently material world is simply a modulation of consciousness itself.[24] This perspective supports a spiritual worldview found at the core of many religious traditions.

This visual representation of the consciousness-centric framework is of course an approximation. The nature of reality is likely far more complex than what can be shown in a two-dimensional image. However, adopting a framework like this can be a useful exercise because it can guide us directionally. And it is from this lens that a deeper perspective about our place in the cosmos can be explored.

CHAPTER 8
SPIRITUAL WAR

The consciousness-centered form of metaphysics discussed in the previous chapter inherently points us toward a view of reality that one might call "spiritual." However, mainstream science largely steers us in the opposite direction. Thus, we're left in a position where humanity is in a great state of ignorance. And this level of ignorance is difficult to rationalize as a mere innocent coincidence. It seems that there are dark and light forces that steer our perceptions, which has big implications for thinking about our place in the cosmos. In this chapter, I delve into these matters.

Social Engineering

The mainstream form of metaphysics that emphasizes matter, rather than consciousness, implies a nihilistic worldview. It suggests that we're part of a random evolutionary process that took billions of years. When we die, that's the end of our consciousness. Life, from this perspective, doesn't have any fundamental meaning beyond whatever one makes up in one's own mind. There isn't meaning built into the fabric of reality. Therefore, from this lens, spiritual or religious concepts are mere superstitions that don't describe how reality actually works. This type of mentality is very much in alignment with the Copernican Principle, wherein "we are not special."

If there exists an agenda to deceive the general population, then it would make sense to direct people toward a matter-focused form of metaphysics rather than a consciousness-focused one. This would create an underlying energy of nihilism rather than one in which humans are inherently important spiritual beings. Arguably, that's exactly what has happened in modern society.

It's perhaps not surprising, then, that scientists who study consciousness "anomalies" face major hurdles. Although some brave scientists have broken through after years of an uphill battle, the landscape continues to be challenging. Psychologist Imants Barušs and cognitive neuroscientist Julia Mossbridge explain the dynamic in their book *Transcendent Mind: Rethinking the Science of Consciousness* (2016): "As a result of studying anomalous phenomena or challenging materialism, scientists may have been ridiculed for doing their work, been prohibited from supervising student theses, been unable to obtain funding from traditional funding sources, been unable to get papers published in mainstream journals, had their teachings censored, been barred from promotions, and been threatened with removal from tenured positions. Students have reported being afraid to be associated with research into anomalous phenomena for fear of jeopardizing their academic careers. Other students have reported explicit reprisals for questioning materialism, and so on."[1]

Neuroscientist Mario Beauregard, PhD, has faced hurdles studying spiritual phenomena as well. In a 2022 interview on the *Skeptiko* podcast, he said: **"It's part of a social engineering program, clearly."** He recalled meeting a "famous" woman at a neurological institute in Canada:

> **She told me: "As long as I'm alive and I'm controlling [this neurological institute], you'll never do neuroscience studies on spirituality. Never."** So she organized something with all the members of the committees—the ethics research committee and the scientific committees—to prevent me from doing these studies. I had earned a grant from the Templeton Foundation based in the United States, but I was not allowed [to pursue

> the research]. They blocked me....I asked them, "What is the reason?" And when I started to argue [with] them regarding spiritual experiences, they were becoming berserk....And then I knew that it was...a war between two contrasting [and] totally different paradigms: dark and light.[2] [emphasis added]

Steven A. Young, PhD, also conceives of social engineering through a chain of belief systems with which we are indoctrinated. Our brainwashing could be viewed as a form of *sorcery*—deception on a mass scale that's akin to a magic trick, almost as if spells are being cast.

What follows is a rendition of Young's framework of social engineering, which he calls the "five elements of scientism."[3]

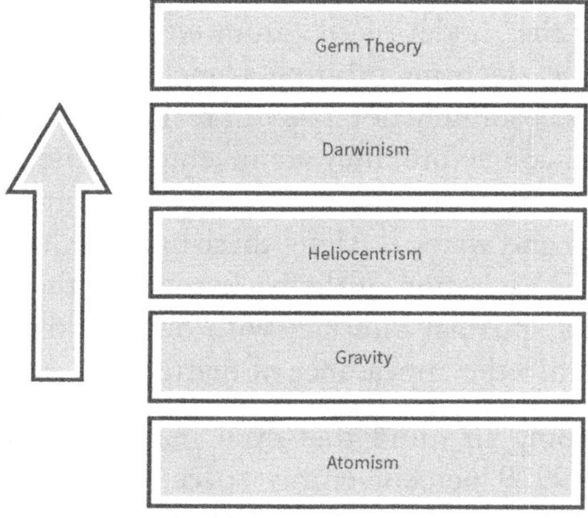

The illustration above is a re-creation of Steven A. Young's framework described in his book *A Fool's Wisdom* (2024). Under this framework, each layer builds on the next to promote new layers of false beliefs, and ultimately results in social engineering.

For Young, it all starts with *atomism*—the belief in tiny units called atoms. In other words, what we call "physical stuff" is comprised of many, many atoms that give the macro-appearance of something physical. This is very much in line with the previously discussed matter-focused metaphysical frameworks that are foundational for modern science. Challenging the atom would indeed challenge much of modern scientific thinking—about the cosmos and otherwise.

The inception of the atom is often attributed to the Greek philosopher Democritus (c. 460–370 BCE); and subatomic particles (protons, neutrons, electrons) were conceived in the late 19th and early 20th centuries (via J.J. Thompson and Ernest Rutherford's experiments).

However, when we descend into the microscopic realm, inferences ultimately have to be made. Young and others are starting to question whether protons, neutrons, and electrons are what we think; or if, instead, we're dealing with things that are more "energetic." Indeed, scientists do observe effects in experiments, but perhaps those effects have been misinterpreted—leading us to a model of "matter" that is likewise incorrect or imprecise. As Young says: "It's clear that the model of the atom we are taught at school, the one with the 'electrons' orbiting a 'nucleus' made of 'neutrons' and 'protons,' is absolutely not a fact of reality. Electrons, protons, neutrons are all conjecture; they are invented in the human mind and given as technical explanations for the elemental phenomena of fire, air, earth and water."[4] Along these lines, perhaps "matter" is simply a specific excitation of the background aether.[5] The aether thus serves as a spiritual interface with our world, transforming consciousness into the appearance of matter.

It's worth keeping in mind that even mainstream physics says matter is 99.99999 percent empty space. That leaves the door open for a tiny bit of solid stuff. But what if even that tiny bit of solid stuff *isn't there*? Then nothing would be truly solid.

This idea works well within a consciousness-centric form of metaphysics. After all, we cannot validate the existence of anything "material" outside of consciousness. Thus, matter is always experienced *within* consciousness. **Anything we touch that's physical, for example, is ultimately just a bunch of sensations that we experience *in our consciousness*; then we infer that it's a physical object made of matter that would exist even if we weren't there to experience it.** This can be felt more acutely when closing one's eyes and feeling one's body against a chair. If all preconceptions and labels are removed, it's evident that the only things experienced are sensations that are registered *in consciousness*. Calling anything physical is an interpretation after mental gymnastics are applied.

German physicist Hans-Peter Dürr put it well when he stated: "Matter is not made of matter."[6] Physicist Fritjof Capra similarly said: "Atoms consist of particles and these particles are not made of any material stuff."[7]

Consider the cosmological implications. Earth is supposedly made of "matter." But if "matter" isn't real in the way we think it is, then how can we really determine the nature of where we live? This wrinkle is often lost in the binary "Globe" or "Flat" debate.

But the implications are even broader, according to Young. The erroneous belief in the atom has a chain reaction that leads to a series of many, many other erroneous beliefs.

First, atomism naturally allows for the belief in gravity. That's because gravity relies on the belief that *mass* has some force within it that attracts other mass. Put another way: matter attracts matter. But gravity doesn't make sense if matter doesn't truly exist.

Young sees the flawed notion of gravity as a precursor to the belief in heliocentrism. Under this cosmology, a "Big Bang" randomly started the universe and filled it with matter, leading to the development of "solar systems," such as the one in which Earth is said to reside. In this solar system, Earth revolves around the Sun. This is similar to what's portrayed in depictions of the atom—a nucleus surrounded by a cloud of electrons. Yet both models seem to be problematic at their core.

Next, Young sees heliocentrism leading to the belief in Darwinian evolution. In other words, heliocentrism includes the billions of years needed for humans to randomly emerge. From this perspective, our existence is not the product of any higher intelligence. We're simply here because of "natural selection"—that is, the survival of the fittest.

There are, in fact, many academic thinkers who've been challenging the science of Darwinism, instead arguing for some form of "intelligent design." They point to anomalies like unexplained complexity, insufficient fossil records, and genetic "missing links."[8]

In fact, Darwinism relies heavily on the notion of random genetic mutations. But "randomness" would need to be mathematically evaluated relative to a "complete genetic dataset," including the "first genes." And since we don't have this data, it's mathematically impossible to conclude that genetic mutations are random. Therefore, the notion of "random mutations"—a core element of Darwinian evolution—is based on inference and theoretical models.[9] (Note: The concept of "natural selection" can still exist to some degree without strict Darwinism. Certain traits indeed give organisms a higher likelihood of survival and reproduction. But the question is whether that mechanism is sufficient to explain the alleged progression of a single-celled organism to complex humans with brains.)

The final part of the chain in Young's framework is that Darwinism leads to a belief in germ theory—because under Darwinism, germs must evolve and adapt in order to survive. Moreover, their pathogenicity—that is, their ability to cause disease—is embedded within this Darwinian perspective. Simply put, germ theory is the notion that microbes *cause disease* by attacking healthy organisms. Thus, microbes are invisible enemies that humans should fear.

An alternative possibility is that germs are more like firefighters who put out fires. Those fires might have arisen from other types of toxicity (whether physical or psychospiritual), but microbes get falsely accused since they're found at the crime scene when in fact they're just doing their job to help the organism heal. Thus, the notion of a "bacterial infection" might be a misnomer. Perhaps the bacteria arrive in large quantities because they act as a "cleanup crew": they assist with the breakdown of dead and dying tissue that was caused by some other trauma or form of toxicity. Certainly, dangerous quantities of bacteria can result (as they try to assist), but the "root cause" of the problem was something other than the germ.

"Viruses" raise separate challenges within germ theory because of questions about their "isolation." Their existence, as replication-competent intracellular parasites, has come under fire because of the way virology experiments are conducted. More specifically, virology is being criticized for employing studies that lack a truly independent variable and proper controls, thus rendering the field one of pseudoscience (by definition). Many Freedom of Information requests have been submitted to health agencies around the world, including the CDC, and repeatedly they fail to provide evidence of a truly purified "viral" sample.[10] Virus skepticism gained traction during the COVID-19 era, but similar questions also arose around HIV (for example, see the Perth Group's paper "HIV—a virus like no other"[11]). Stefan Lanka, a classically trained virologist, has also been a leading critic of the field of virology.

Reevaluating germ theory is vital because it can help us reveal the true determinants of health and disease. In particular, we might then be inclined to take more personal responsibility for our health—and consider the link between consciousness and our physical symptoms (see, for example, German New Medicine[12]). Such a shift could lead to a world in which humans have greater freedom and independence from authority structures. During the COVID-19 era, we experienced firsthand the ways in which germ theory can be weaponized to control the population. In fact, the World Economic Forum's vision for society, known as the "Great Reset," was initiated because of COVID-19. The book *COVID-19: The Great Reset* (2020) explicitly laid out this vision, which included centralizing power (among many other concerning ideas discussed in my book *An End to the Upside Down Reset*).

Germ theory also makes way for vaccinations, which are mechanisms by which power structures can gain access to the human body via direct injections. They also create an ability to control the population, whereby one's ability to work or travel could be contingent upon receiving a vaccination.

The scientific flaws of germ theory are complex, and I've only presented a snapshot here. Further detail is included in my book *An End to Upside Medicine*; and in foundational works such as

A Farwell to Virology by Dr. Mark Bailey; *Virus Mania* by Torsten Engelbrecht et al.; *What Really Makes You Ill* by Dawn Lester and David Parker; *The Contagion Myth* by Dr. Thomas Cowan and Sally Fallon Morell; *Breaking the Spell* by Dr. Thomas Cowan; *The Final Pandemic* by Drs. Samantha and Mark Bailey; *Can You Catch a Cold?* by Daniel Roytas; Mike Stone's ViroLIEgy.com; "The End of COVID" series produced by Alec Zeck, Mike Winner, Kelly Brogan, MD, and others, available at https://theendofcovid.com/; *The Viral Delusion* documentary series, directed by Michael Wallach; *TERRAIN: The Film* directed by Marcelina Cravat and Andrew Kaufman, MD; and Freedom of Information requests submitted by Christine Massey and her colleagues, which are accessible at https://www.fluoridefreepeel.ca/.

Arguably all of the beliefs laid out by Young—atomism, gravity, heliocentrism, Darwinism, and germ theory—are directing humanity's consciousness away from the truth. Instead, they direct mass consciousness toward a nihilistic and fear-based mindset that is much more easily controllable than a human believing that he or she is innately powerful, important, and free.

Dark Metaphysical Forces

If Young's framework is even directionally accurate, and if even a fraction of the perspectives shared in this book are true, they are hard to rationalize as mere incompetence. In other words, it's hard to conceive that our drift into false paradigms is purely the result of innocent mistakes.

Perhaps false paradigms are being propagated intentionally. Not many people would be needed to enact such an agenda if they can use blackmail, bribery, threats, and mind control to manipulate key individuals in society.

But *why* would such social engineering be important? What would entice people to brainwash the public? And how could such an agenda remain over long periods of time if people have relatively short lifespans?

The answers might lie in the domain of consciousness—realms that transcend purely physical considerations. **Perhaps there are nonphysical, spiritual forces that have much longer time horizons than humans do, and they work through humans in order to enact their goals.** Why they might want to do so is open for speculation, but a common theory is that dark forces literally feed off of human energy—in particular, human suffering.

Within a consciousness-centric metaphysical framework, the existence of such forces is plausible. For instance, the famed computer scientist Alan Turing stated in 1950 that the statistical evidence for telepathy is "overwhelming," which led him to then state: "It does not seem a very big step to believe in ghosts and bogies."[13] Put another way, once we accept that consciousness is not localized to, or dependent on, a brain, then it's possible for forms of consciousness to exist that don't have physical bodies. Broadly speaking, they're often referred to as *spirits*. Maybe they even have an ability to take on different forms (known as *shapeshifting*).[14]

Along these lines, Helané Wabheh's book *The Science of Channeling* (2021) compiles scientific evidence for the ability of humans to "tune in to" some of these nonphysical intelligences.[15] Sometimes this can occur through voluntary invocations, but it also seems possible to tune in without being aware of it. If our brain has an antenna-like quality, then this is possible. After all, we don't really know where our thoughts originate. It might be the case that an individual's consciousness is constantly being influenced without the person's explicit knowledge. Therefore, the ability to manage emotions and beliefs seems critical.

In my book *An End to Upside Down Contact* (2022), I document a wide variety of encounters with nonphysical forces, some seemingly benevolent and others not. The focus of the forthcoming discussion is on the darker forces. This focus is necessary because of the unquestionably deceptive nature of our reality. However, it's important to acknowledge that benevolent forces seem to be real too. Practically speaking, if everything in the world were purely evil, it's implausible that we'd even have the freedom to question paradigms. Additionally, our world is filled with loving acts, which is suggestive of a benevolent spiritual force. Likewise, there are many otherworldly experiences, reported historically and in modern times, in which people have direct mystical encounters with seemingly benevolent beings—often reporting indescribable states of "unconditional love" and even being miraculously healed. Preston Dennett gives many examples in his 2019 book, *The Healing Power of UFOs: 300 True Accounts of People Healed by Extraterrestrials*.[16]

Or consider Anita Moorjani's case. She was healed after having a transformative near-death experience in 2006. She was in a coma during a bout with terminal cancer and has since tried to put into words the shift that her consciousness underwent: "What I can only describe as superb and glorious unconditional love surrounded me, wrapping me tight as I continued to let go. It didn't feel as though I had physically gone somewhere else—it was more as though I'd awakened, perhaps finally being roused from a bad dream. My soul was finally realizing its true magnificence. And in doing so, it was expanding beyond my body and this physical world."[17] Her doctors were stunned when her tumors disappeared following the resuscitation.

Regarding darker forces, consider the work of Richard Gallagher, MD. He is a Princeton and Yale graduate and a professor of clinical psychiatry at New York Medical College who has personal experience with patients suffering from "demonic possession." He details such accounts in his book, *Demonic Foes: My Twenty-Five Years as a Psychiatrist Investigating Possessions, Diabolic Attacks, and the Paranormal* (2020). He states his conclusion: **"A segment of [the] invisible world seems to be mysteriously but remarkably hostile to human beings and seeks their physical and spiritual destruction. On rare occasions, like some kind of cosmic terrorist, that segment shows its true colors."**[18] [emphasis added]

In fact, there are some instances in which individuals intentionally summon such dark forces. Earlier this was alluded to in the discussion of the "sex magick" ritual engaged by Jack Parsons (one of the early influences at NASA), with assistance from

L. Ron Hubbard (the founder of the Church of Scientology). One has to wonder if this sort of ritual was just the tip of the iceberg. *The Telegraph* summarized the situation well in its 2017 article titled "Sex, rocket science, and Satanism: Meet Nasa's real hidden figures."[19]

There is, in fact, a long history of secretive, heinous, and sophisticated rituals that date back to biblical times. They have also been described in different forms by the Aztecs and Egyptians, among others. The intent of many such rituals is to call in dark spirits, likely for the purpose of obtaining power.[20]

The rituals give vital insights into the psychology of evil and might help us understand the drive behind deception in our world. Only minor snippets will be covered here because the details are so unpleasant.

Although many instances of ritualistic abuse are hidden, *some* cases have been uncovered over time. In fact, they are discussed explicitly in state laws in Idaho and Illinois, and there are clear patterns to the activities. What follows is an excerpt from a law in Idaho (1990), and the law in Illinois is similar:

> 18-1506A. RITUALIZED ABUSE OF A CHILD — EXCLUSIONS —PENALTIES — DEFINITION.
>
> (1) A person is guilty of a felony when he commits any of the following acts with, upon, or in the presence of a child as part of a ceremony, rite or any similar observance:
>
> > (a) Actually or in simulation, tortures, mutilates or sacrifices any warm-blooded animal or human being;
> >
> > (b) Forces ingestion, injection or other application of any narcotic, drug, hallucinogen or anaesthetic for the purpose of dulling sensitivity, cognition, recollection of, or resistance to any criminal activity;
> >
> > (c) Forces ingestion, or external application, of human or animal urine, feces, flesh, blood, bones, body secretions, nonprescribed drugs or chemical compounds;

(d) Involves the child in a mock, unauthorized or unlawful marriage ceremony with another person or representation of any force or deity, followed by sexual contact with the child;

(e) Places a living child into a coffin or open grave containing a human corpse or remains;

(f) Threatens death or serious harm to a child, his parents, family, pets or friends which instills a well-founded fear in the child that the threat will be carried out; or

(g) Unlawfully dissects, mutilates, or incinerates a human corpse.[21]

Along these lines, Anne Johnson Davis's autobiographical book *Hell Minus One: My Story of Deliverance from Satanic Ritual Abuse and My Journey to Freedom* (2008) describes her experiences as a victim as a young girl. The foreword of the book was written by a Utah detective from the state attorney general's office who investigated that case. He confirmed that the perpetrators wrote confession letters detailing their crimes and added, "They confessed to us—in person. The allegations were confirmed."[22]

This is significant because such allegations are often dismissed as products of "false memories." Yet this is a validated case. In other instances, it's indeed difficult to know what is real, particularly because the rituals often involve drugging and forms of torture designed to suppress memories. But in Davis's case, her memories came flooding back later in life (which isn't uncommon), and through unique circumstances, she was able to confront the perpetrators who decided to confess. Often the abused individual might just be considered crazy, with no way to validate the memories.

The bigger point here is this: If even some cases are real, then we can gain insights into the broader spiritual war that we all might be involved in (whether we know it or not).

In Davis's case, she recalls seeing dark spirits entering into the ritual.[23] Jay Parker endured similar abuse as a child, including incest, and he experienced dark energies directly. When interviewed by filmmaker Sean Stone, he recalled that a dark intelligence seemed to possess his parents while they were engaging in nefarious acts. Parker says, **"My father's eyes would flash black, solid black,"** as would his mother's.[24] [emphasis added]

Along similar lines, a *60 Minutes Australia* episode that aired in 1989 featured a young British girl who was subjected to ritual abuse after being brought into the circle by her own grandmother. A psychiatrist validated her experiences as real and not the product of a psychotic or imaginative mind. Clearly traumatized (almost despondent and emotionless), she spoke about being brought to a place in the English countryside that looked like a "castle." She said the route to get there was intentionally hidden from the victims. She also noted that the people there were "very rich" and had an ability to destroy evidence of murders with chemicals. They also referred to "Lucifer," as she recalled. She detailed the horrors to which she was subjected, including being raped, being forced to have an abortion, and then being forced to eat the fetus.[25]

Survivors of such abuse often describe links to powerful people—even governments. Anneke Lucas, for instance, was sold by her mother into a murderous pedophile ring in Belgium with "VIPs" and people "on the world stage." Ritual abuse was involved as well.[26] She documents her accounts in her 2022 book *Quest for Love: Memoir of a Child Sex Slave*. Another example is Serena Faith Masterson, author of *I Am Serena* (2020), who was born into a satanic cult and sold into the US government's mind-control program. As a result of this torture, she developed dissociative identity disorder, including over 300 unique personalities. Only through a rare and miraculous series of events was she able to heal and tell her story. Her message is ultimately uplifting: she feels strongly that her soul grew from the experience and that she is able to contribute to the healing of others as a result. She seems to imply that suffering, while horrific in many ways, can paradoxically lead to healing and growth.

There are many cases that show a link between dark metaphysical forces and power structures in our world. Sean Stone's documentary *Best Kept Secret* (2021), and John DeCamp's book *The*

Franklin Cover-Up: Child Abuse, Satanism, and Murder in Nebraska (1992) provide additional insights, including links with child-sex trafficking and child pornography. Another important resource that validates the reality of ritual abuse is a more than 500-page compendium featuring essays from more than twenty-five contributors, titled *Ritual Abuse in the Twenty-First Century: Psychological, Forensic, Social, and Political Considerations* (2008).

The point of acknowledging this dark underbelly of our world is not to dwell on evil but rather to acknowledge reality. When we ask, *Why is there so much deception in our world? What is driving this?*, we're forced to consider these deeper energies that might be behind it.

Parker sums it up by saying that "the suffering and dissatisfaction of the human soul is what feeds the [dark] spirits....It's all an energy game....We are cattle for them, and we feed them." Yet, in spite of the unspeakable abuse he endured, he still feels that the benevolent force is greater: "Divine love is much more powerful than anything else."[27] [emphasis added]

Contact in the Cosmos

The notion of a spiritual war raises broader questions about our place in the cosmos relative to advanced life-forms. That naturally introduces questions about "UFOs" and their potential role in all of this. Are any life-forms coming from distant locations to meddle in human affairs? Are they players in the spiritual war?

From a Globe Skeptic perspective, one place to explore would be the lands "beyond the Antarctic ice well." As discussed in chapter 4, there have been (unverified) claims of other life in those lands. The notion of "extraterrestrials," from this lens, could simply refer to intelligences that come from the "extra-territories"—lands that exist beyond the ice-wall enclosure of the Flat Earth.[28]

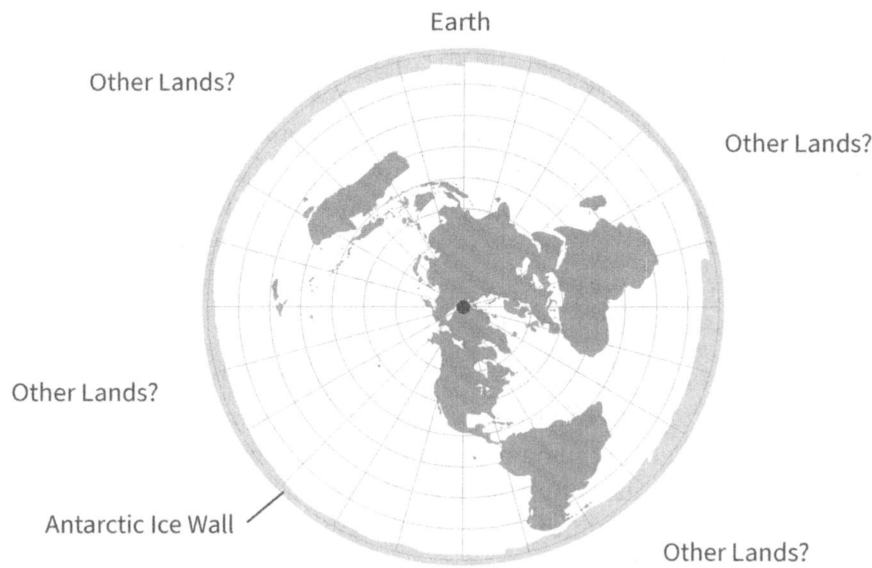

The Flat Earth framework leaves open the possibility of other lands that exist beyond the Antarctic ice wall surrounding the "lake" of Earth. From this lens, the notion of extraterrestrials might refer to intelligent beings that exist in these other lands—the "extra territories." However, restrictions imposed by the 1959 Antarctic Treaty limit truly free travel beyond the 60th S latitude. Notions of these extra territories, and beings that might live there, remain speculation at this point.

Indeed, there are many instances in which cultures have described interactions with beings from *somewhere* else. For instance, Montana State University professor Ardy Sixkiller Clarke has studied Native American and Mesoamerican cultures who speak of the "Star People" and the "Sky People."[29] But perhaps it is not known with certainty where they came from or what technology they used to travel—multidimensional or otherwise. After all, a US government document declassified in 2024 discusses a project that mentioned "physical transport across dimensional/space-time barrier" and "de/re-materialization techniques."[30]

From a Globe Skeptic lens, the possibility of a firmament enclosure might preclude certain forms of physical travel. But perhaps metaphysical travel—via consciousness—is possible. In fact, mystic Emanuel Swedenborg wrote a book titled *Other Planets* (1758) in which he had otherworldly experiences with his consciousness alone. He explained the process: "Angels who had been sent by the Lord took me to a particular extrasolar planet, where I was allowed to observe the planet itself....**Being taken to other planets in the universe does not mean being taken there in the body, but rather in the spirit, which is led through changes of the state of its inner life that seem like movement through space.**...As a result, a person's spirit can be transported and brought to distant places while the person's body remains in the same place."[31] [emphasis added]

On the other hand, there are many contemporary allegations of a Secret Space Program (SSP) in which people claim to have traveled physically to other planets, the Moon, and beyond. From this lens, a faked Moon landing could be attributed to, say, secret conflicts with aliens encountered on the Moon that the government wants to keep quiet. SSP experiencers also speak about the traumatic mind control they endured, which naturally raises questions about their memories. The SSP is covered extensively by Michael Salla, PhD, in his books and on his Exopolitics platform.[32] (Note: Most SSP survivors tend to say that Earth is spherical. However, one survivor said in 2016 that he doesn't know if Earth is flat or a sphere: "I've seen a lot of things, and it's hard to describe—even seeing it [myself]—because I can't tell you whether [the way Earth looked was] being played into my mind to make it look that way. I haven't gone out in my own craft, with my friends, where we're not tricking each other, and then seen [Earth]....Whether it's flat or round, but we're in a type of mental, spiritual, physical plane—and it reacts to how our consciousness perceives and expects it—then [Earth's shape] is either/or. It's whichever you look at it as. And it's way more complex than that."[33])

Moreover, perhaps looking "out there" for intelligent life causes us to downplay the possibility of life-forms that are much closer to us than we realize. Leading UFO researcher Jacques Vallee remarks: "The simple truth is this: if there is a form of life and consciousness that operates on properties of space-time we have not yet discovered, then it does not have to be extraterrestrial [as in, from outer space]. It could come from any place and any time, even from our own environment...It could...coexist with us and remain undetected. The entities could be multidimensional beyond space-time itself. They could even be fractal beings. The earth could be their home port."[34]

Similarly, many cultures have spoken about life *below* us (including "underworlds").[35] The reality is that we know very little about the ground beneath our feet, or even what's contained within Earth itself (as discussed earlier).

And yet, the public narrative around "contact" with nonhuman intelligence often focuses on faraway lands. This has accelerated since a December 2017 *New York Times* article hit the Sunday front page.[36] But there is growing concern that the UFO narrative could be weaponized to create unnecessary fear. For instance, Vallee writes in his book *Revelations: Alien Contact and Human Deception* (1991): **"Somebody is going to an awful lot of trouble to convince the world that we are threatened by beings from outer space....The idea that we are about to encounter new enemies in outer space contains unfathomable power."**[37] [emphasis added]

In fact, insights from Dr. Carol Roslin further advance Vallee's idea about the weaponization of "space." In 1974, Roslin met Wernher von Braun, the former Nazi SS officer who had a significant role in NASA's early years. They were both working at Fairchild Industries at that time. Von Braun was dying of cancer and passed along messages to her until his death in 1977, telling her about a long-term plan to "weaponize space" by building "space-based weapons." He relayed to her that the "final card" would be to build such defenses against "aliens, extraterrestrials." But, as he told her, "All of it is a lie."[38] Thus, it's often speculated that a fake alien threat could be used by governments to enact authoritarian policies while the public is absorbed in fear. The belief in the Globe model of "outer space" would be central to such an agenda.

However, the UFO phenomenon is highly complex—sometimes more so than it's made out to be. Just because governments could manipulate the narrative doesn't mean that there aren't strange crafts in the sky. In fact, they've been reported throughout history—as detailed in Vallee and Chris Aubeck's 2010 book *Wonders in the Sky: Unexplained Aerial Objects from Antiquity to Modern Times.*

Additionally, there are some bizarre phenomena associated with UFO-craft encounters that transcend the narrative of merely physical phenomena. The "Vallee-Davis model," summarized in the 2003 paper "Incommensurability, Orthodoxy and the Physics of High Strangeness: A 6-layer Model for Anomalous Phenomena," describes six layers often reported among those who have had UFO sighting experiences—only *one* of which is purely physical. The "psychic" layer of experience, for example, can include levitation, premonitions, "personality changes promoting unusual abilities in the witness," "maneuvers of a [UFO] appearing to anticipate the [witnesses'] thoughts," and healing, among other things.[39]

All that said, the notion of government deception in the context of UFOs cannot be ignored. Disinformation has even been planted to propagate false narratives. Paul Bennewitz's case is a cautionary tale: he was fed false information that he then spread widely. As written in Greg Bishop's 2005 book, *Project Beta: The Story of Paul Bennewitz, National Security, and the Creation of the Modern UFO Myth*, "It all started with just a few little, carefully chosen lies."[40]

Government officials are even remembered alongside "alien" creatures in certain allegations of "alien abductions." These are reported so often that they have a special name in the UFO field: MILABS (short for military abductions).[41]

And yet there have been many reports of alien abductions that *don't* involve military officials. This has been studied extensively by John Mack, MD, the former head of psychiatry at Harvard and a Pulitzer Prize winner; and by David Jacobs, PhD, a former associate professor of history at Temple University—and both of them have taken the phenomenon seriously.

Moreover, some abductions allegedly occur on the level of *consciousness*.[42] This is reminiscent of Swedenborg's notion of spiritual travel to other planets. It also aligns with Rick Strassman, MD's research on the psychedelic DMT (dimethyltryptamine) in the 1990s. Participants in his study reported vivid abduction-like encounters with higher intelligences when DMT was administered intravenously, and yet their bodies were physically located with the experimenter. Strassman and many of the participants weren't even familiar with the alien-abduction phenomenon at the time, and yet the reports closely aligned with the existing literature elsewhere.[43]

Encounters with advanced intelligences have been reported to include a wide variety of species ranging from "greys" to reptilians to insectoids, and more. Complicating matters further is that advanced beings are described as having an ability to change their appearance (known as *shapeshifting*) and can alter our memory of them (by creating in our consciousness "screen memories"). **Thus, it can be challenging to know if a being has benevolent intentions or if it is a deceptive trickster—a "wolf in sheep's clothing" representing false light.[44]**

Reports of contact with nonhuman intelligence ultimately raise more questions than they answer. In fact, there is a special term used to describe the wide variety of bizarre reports: *high strangeness*. What seems apparent, however, is that we are not alone in the cosmos. How these beings interface with humanity, who they are, what they want, and how they're involved in the spiritual war is much less clear.

What Is This Place?

The notion of a spiritual war also raises questions about human origins. Who or what put us here? Are we part of a benevolent creation that allows our souls to have experiences and evolve through reincarnation cycles? Are there beings who want to keep our souls trapped here? If a "firmament" exists, is it a form of entrapment? Are we enslaved, unknowingly serving our masters, or are we free?

One place to look for answers is ancient scripture. **But even that is a difficult exercise because there are so many interpretations and variations on the stories.**

For instance, consider the work of Mauro Biglino, an Italian biblical translator and scholar. According to his bio, he "has directed and supervised the translation and publication of 17 books of the Old Testament for Edizioni San Paolo, Italy's main Catholic publisher, whose Vatican-approved texts are used in theological, ancient Hebrew, and biblical studies schools and universities."[45]

In his book *Gods of the Bible: A New Interpretation of the Bible Reveals the Oldest Secret In History* (2023), Biglino discusses the possible meaning of the word *Elohim* as it relates to human origins. Traditionally this word been thought of as synonymous with *God*. But his analysis leads him to conclude that perhaps the term refers to a *group* of beings rather than a single being. Biglino writes:

> The biblical term "Elohim," did not refer to one spiritual, transcendent, omniscient, and omnipotent "God" but to many flesh-and-blood individuals....
>
> The Elohim lived long enough to be considered immortal, even though they were not....
>
> They were individuals who traveled on flying machines called [ruach], [kavod], [merkavah], and [cherubim]....
>
> The Elohim were never considered "gods" in the conventional sense. They were objects of respect and inspired terror simply because of their great power, guaranteed by the technology they possessed....
>
> Their goal was to define the power structures in the various territories where different civilizations were developing....
>
> The Elohim knew the laws of nature and the cosmos and passed them to their most devout followers, creating castes of kings/governors/priests: the so-called "initiates."[46]

Furthermore, he concludes that "Yahweh" from the Bible, regarded in traditional religion as "God," was "only one of [the Elohim]."[47] This perspective doesn't deny the existence of a transcendent force, but rather suggests that the transcendent force is not exactly as it's been implied by traditional biblical interpretations.

Biglino's general interpretation bears similarities to those of Paul Wallis, a former archdeacon in the Australian Anglican Church. His *Eden* series of books examines such alternative narratives of our origins and shows similarities between biblical narratives, Sumerian (Mesopotamian) myths, Mayan (Mesoamerican) creation stories, and others from around the world. His analysis of the texts in 2020 suggests to him that humans might have been a literal biological creation of more advanced beings, and that "our mental capacities and powers of perception were downgraded. To maximize our usefulness to our overlords, our brains' default settings were deliberately turned down."[48]

The Nag Hammadi scriptures raise intriguing possibilities as well. These were bound books found in a jar in Egypt in 1945 and were translated into English in 1977. They were believed to have been written around the second and third centuries AD and provide controversial takes on Christianity. Some historians theorize that they were buried long ago for protection. The teachings were likely considered heretical.[49]

Regardless of whether the texts should be taken literally, they raise thought-provoking frameworks—particularly those about human origins. For instance, "The Secret Book of John" starts by describing "The One," which is "a sovereign that has nothing over it" and is "eternal." The One is further identified as "God and Parent, Father of All, the invisible one that is over All, that is incorruptible, that is pure light at which no eye can gaze. The One is the Invisible Spirit."[50] The One spawned a female counterpart, named "Barbelo," who appeared in his presence after he reflected on his image and became "enamored."[51] From their union came a lineage that ultimately led to a rogue being named Yaldabaoth. According to the scripture, "He called himself God."[52] In fact, Yaldabaoth even stated: "I am a jealous god and there is no other god beside me."[53] He and his minions then decided to create humans,[54] but became jealous and then threw their creation "into the lowest part of the whole material realm."[55] That ultimately led to the "imprisonment of humanity."[56]

Another writing within the Nag Hammadi scriptures states: "The rulers threw humanity into great confusion and a life of toil, so that their people might be preoccupied with things of the world and not have time to be occupied with the holy Spirit."[57] [emphasis added]

Although the creation stories in the Nag Hammadi writings do have a conspiratorial undertone, the ending is ultimately positive. **As one of them says: "Light overcomes darkness."**[58] [emphasis added]

These are just a few examples, and a more comprehensive analysis would require looking at many scriptures from cultures all over the world. However, interpreting ancient texts always poses a challenge because of translation nuances. Our modern understanding of life might also cloud our ability to relate to what other cultures were experiencing. Moreover, we don't know the extent to which the writings are literal, figurative, true, partially true, or intentionally deceptive.

On a related note, theories of human origins often lead to explorations of "lost civilizations." Plato spoke of Atlantis—an advanced society that existed "before the great flood." The work of Graham Hancock and others similarly points to the possibility of an advanced society that existed thousands of years ago. Göbekli Tepe in Turkey is a commonly referenced archaeological data point because it suggests that an advanced civilization existed much earlier than modern textbooks say it should.[59] There is also a growing discussion about the possibility that an advanced civilization was here as recently as the 1800s (known as "Tartaria"), but society was "reset" such that we're unaware of this recent history.[60] Furthermore, Mike Wilkerson has explored geological (petrification) evidence potentially suggesting that our world had gigantic humans, animals, and trees—which harkens back to ancient scriptures and mythologies. He details these explorations on his Stellium7 YouTube channel.[61]

Although we might not have definitive answers to these questions today, they are essential subjects if we want to truly understand who we are and what our place in the cosmos is.

CONCLUSION

FREEDOM THROUGH TRUTH

In this book, I have examined fundamental beliefs that we've been taught and have asked whether they have a solid foundation. Repeatedly, the answer seems to be that there's more to the story.

There is a common tendency to assume that modern science has a basic grasp on the nature of the cosmos and reality more broadly. Because of this assumption, there is often resistance to challenging the paradigm. The thinking is: *We can explain so much of the universe with the current model. It works pretty well.*

Except that…it doesn't.

There exists no unifying theory of everything that reconciles the leading theories of physics (gravity/general relativity and quantum mechanics). And 96 percent of the universe is labeled mysterious dark matter and dark energy. Some physicists are now saying that dark matter has been outright falsified, and dark energy is highly problematic too. The ripple effects of these fundamental problems cannot be emphasized enough. They are like the first dominoes that then put into question all of our basic views about the cosmos and our place in it.

Dark matter and dark energy are needed to preserve the flawed theories of gravity and general relativity; and if those aren't what we think, then the cosmological house of cards starts to crumble. More specifically, the alleged entities of dark matter and dark energy prop up unfounded confidence in a Big Bang that occurred 13.8 billion years ago, which allegedly created our heliocentric solar system, wherein humans randomly evolved and now exist as a tiny speck in a vast universe. What if this whole narrative is based on false premises?

Meanwhile, the Copernican Principle pervades modern thinking, teaching us that we aren't special; and yet the geocentric, stationary alternative is arguably equivalent, if not scientifically superior. Even the nature of where we live has legitimate reason to be reexamined, as does the nature of what it means to be human.

The irony is that many who defend the mainstream model lack an understanding of its fundamental flaws, and yet they religiously defend it.

So, what are we to do?

A good place to start is shifting one's own consciousness. That entails orienting one's mind toward a desire to understand the truth about reality. The process can involve identifying what is *not* true in an effort to narrow down the possibilities of what *is* true. It means not being satisfied with falsehoods and deceptions.

The impact of such a mentality is profound when considered within the context of a consciousness-centered form of metaphysics. Reality, from this lens, has a "malleable" quality. *How* this might work is unclear, but the effect has been demonstrated scientifically (at a small but highly statistically significant level).

For instance, experiments on psychokinesis (mind-matter interactions) suggest that the human mind is able to alter the behavior of machines (known as random-number generators [RNGs]). To be clear, humans influence the way the machines behave *without touching the machines*. The effect is seen when participants are asked to apply a mental *intention* toward altering the machines. This has been studied by researchers such as Princeton University's former dean of Engineering, Dr. Robert Jahn, and his team at the Princeton Engineering Anomalies Research Lab (which operated from 1979 to 2007).[1]

This effect is also seen when RNGs are set up all over the world, as studied within the Global Consciousness Project—an offshoot of the work done at Princeton. The machines behave nonrandomly when there are major world events that seem to capture the attention and/or emotional strings of the collective consciousness (such as 9/11). Thus, the results suggest that collective consciousness impacts the behavior of machines. Most people aren't even aware that the machines exist. Yet the machines behave differently when many people shift their consciousness. The power of the mind is profound; it seems to be able to literally shift reality.[2]

On a more speculative note, the "Mandela Effect" deserves consideration. This refers to apparent "glitches in the matrix" whereby a former truth is now remembered as something different. As stated by Cynthia Sue Larson, who holds a degree in physics from UC Berkeley and wrote *The Mandela Effect and Its Society* (2024), the Mandela Effect "has to do with groups of people remembering something differently than what the official description of historical facts might be."[3] So this is another possible instance in which reality seems to be malleable. For instance, the famous *Star Wars* line often remembered as "Luke, I am your father," is actually "No, I am your father."

Another commonly cited example relates to airplane engines that now have a different location than what is often remembered.[4] I was even told this directly by a former pilot who said that the location changed. Many other examples have been compiled, even Bible verses that have allegedly shifted. Of course, it's difficult to know if they are instances of basic "misremembering" or whether reality itself actually changed.[5] But if this phenomenon is in any way real, one might wonder what the implications are for where we live: Could it be that Earth is, and has been, shifting its form, thereby further complicating the Flat Earth versus Globe debate?

Similarly, this topic raises questions about causality. The typical view of causality is that time moves sequentially from the past to the present to the future. But some scientific phenomena suggest that the future can impact the past. Precognition (sometimes referred to as *presentiment*) is one such phenomenon that's been studied by Cornell University's Daryl Bem, cognitive neuroscientist Julia Mossbridge, and others. In these studies, participants view a computer screen and are shown images that are either peaceful or arousing (such as a violent or erotic image). The sequence of the images is *randomly* generated by a computer, so no one—including the participants—should know what type of image will appear next. And yet the researchers find that participants' bodies subtly respond to the images *before* the image is presented. For instance, an arousing image causes more of a bodily response before the image appears on the screen, as measured by physiological indicators in the eyes, skin, brain, and heart.

The results are small, but highly statistically significant. They suggest that there is an apparently "psychic" effect, and the results have appeared in peer-reviewed journals.[6] Similarly, there are many reports of precognitive dreams in which people accurately dream about something before it happens.[7]

Could it be that time is not linear in the way we interpret it to be? After all, in our experience, we only know the present moment to be true. The "past" is a mere memory that occurs in the present moment, and the "future" is a thought in the present moment. The implications of altered causality are profound: Might we be able to positively impact the past by shifting our consciousness in the present moment?

Regardless of such speculation, perhaps the malleability of our world can be utilized for targeted, practical purposes. For example, there are cases in which group meditation seems to alter reality in a positive way. These are documented reports of "a decrease in negative social trends, such as a decline in war fatalities, and... an increase in cooperation between nations" when groups of people meditate. **The collective orientation of consciousness is thus tied to positive changes in the physical world.** This phenomenon, known as "the Maharishi Effect," has been validated by twenty peer-reviewed studies explored in Barry Spivack and

Patricia Anne Saunders's book *An Antidote to Violence* (2020).[8]

What might the implications be as we collectively begin to lose tolerance for falsehoods? What will happen when we let go of belief systems that make us feel limited? In essence, the process is one of *unlearning*. But it entails being proactive rather than passive.

The ultimate outcome is perhaps a liberated mind, which is much less susceptible to control by dark forces within our apparent spiritual war. One might even wonder if the consequences of a liberated mind extend into the spiritual realm. Are there benefits for the soul that persist beyond this body and beyond this life?

Thus, the journey is inward at its core. Trying to change the power structures of our world is perhaps an unrealistic goal, whereas what we can readily control is our internal state of being—our individualized consciousness. We can *choose* to orient our consciousness in the direction of truth and hold much more loosely to beliefs that haven't been investigated. Ultimately, this mindset is less about "fighting the external system" and more about "cleaning up the internal system."

Similarly, trying to change others is a futile exercise. People shift on their own time frames. Offering assistance and sharing information is, of course, a compassionate act, but forcing people to engage and alter their views rarely ends well.

By elevating our own consciousness, perhaps there are ripple effects that benefit the world in unseen ways. When we acknowledge our spiritual essence—that we're innately free beings, and that we're all interconnected—we naturally exude a fearless liberation that is contagious. It's an energy.

If we are truly like "whirlpools in a stream," then it follows that a shift in one part of the stream necessarily alters the entire stream. Therefore, even small shifts in one person's consciousness could theoretically have a nonlinear, or even exponential, impact on the collective (sometimes known as *the butterfly effect*). But the benefits occur as a natural by-product of one's own inner work and self-liberation. Thus, we help ourselves and the world simultaneously—a true act of interconnectedness and service.

By overcoming our own internal blockages, we open the possibility for others to do the same. It's akin to the phenomenon in which the first four-minute-mile record was broken by a runner in 1954—after which, many, many runners were able to break it again. Maybe our world will become an unimaginably glorious place once we start breaking psychological barriers imposed by the falsehoods we've been collectively conditioned to believe.

Ultimately, society is becoming increasingly bifurcated, wherein each of us seems to be faced with a binary choice: we can either accept the narratives imposed upon us with blind faith, or we can examine assumptions and think for ourselves, acting with intellectual humility along the way.

Which will *you* choose?

ACKNOWLEDGMENTS

First, I want to thank my publisher, Waterside Productions, for once again supporting me in my journey to uncover the truth (wherever it may lead me). I give immense thanks to Gayle Gladstone for her ongoing support, and to the late Bill Gladstone, whom I greatly miss (though his support in spirit is apparent). I also thank those who have assisted me in the process of publishing this book: Jill Kramer (copyediting), Joel Chamberlain (design and digital art), Ken Fraser (cover art), Jennifer Uram (contract support), Sandi Schroeder (index), and Justin Levy (audiobook).

Thanks to Steve Young, PhD, Dave Weiss, and Niraj Mehta for reviewing the manuscript and providing feedback; and to Natasha Nazerali, Meli Neubek, Kelly Brogan, Alec Zeck, Kathryn Ducey, Edith Ubuntu Chan, Dawn Lester, Eileen McKusick, Polla Pratt, Sam Wu, Nancy Heydemann, Callie Kron, Rochel Leah Bernstein, Mark Edmond, Brandon Thomas, Tim Newton, Dori Ben-David, Michael Polansky, Dani Antman, Tim Rothschild, John Paul Rice, Yerasimos Stilianessis, Joel Rafidi, Rusty Shores, Deepak Ramapriyan, Oz Sozen, Alec Day, Susan McIntosh, John Chavez, Katherine Housh, and Tiffini Jacobs for their various forms of support during the period in which I researched and wrote this book. Finally, I thank my family and friends for their consistent encouragement and feedback.

ACKNOWLEDGMENTS

ENDNOTES

OPENING QUOTATION

1 Dr. Tom Cowan, "Q&A/ 'Do Viruses Exist?' Article Webinar from May 29th, 2024," https://rumble.com/v4y9xlr-q-and-a-do-viruses-exist-article-webinar-from-may-29th-2024.html?, 4:30–6:30 (quotation patched together without using ellipses).

INTRODUCTION: ERRORS IN PHYSICS AND THINKING

1 Moskowitz, "What's 96 Percent of the Universe Made Of? Astronomers Don't Know," https://www.space.com/11642-dark-matter-dark-energy-4-percent-universe-panek.html.

2 As noted by Gianfranco Bertone and Dan Hooper in their 2018 paper, *A History of Dark Matter*: "The Swiss-American astronomer Fritz Zwicky is arguably the most famous and widely cited pioneer in the field of dark matter. In 1933, he studied the redshifts of various galaxy clusters, as published by Edwin Hubble and Milton Humason (Hubble and Humason, 1931), and noticed a large scatter in the apparent velocities of eight galaxies within the Coma Cluster, with differences that exceeded 2,000 km/s (Zwicky, 1933)....Zwicky started by estimating the total mass of Coma to be the product of the number of observed galaxies, 800, and the average mass of a galaxy, which he took to be 109 solar masses, as suggested by Hubble. He then adopted an estimate for the physical size of the system, which he took to be around 106 light-years, in order to determine the potential energy of the system. From there, he calculated the average kinetic energy and finally a velocity dispersion. He found that 800 galaxies of 109 solar masses in a sphere of

106 light-years should exhibit a velocity dispersion of 80 km/s. In contrast, the observed average velocity dispersion along the line-of-sight was approximately 1000 km/s. From this comparison, he concluded (Zwicky, 1933): 'If this would be confirmed, we would get the surprising result that dark matter is present in much greater amount than luminous matter.' This sentence is sometimes cited in the literature as the first usage of the phrase 'dark matter.' It is not…and it is not even the first time that Zwicky used it in a publication." Bertone and Hooper, https://link.aps.org/accepted/10.1103/RevModPhys.90.045002.

3 Yale University, COSINE-100 Dark Matter Experiment, "What is Dark Matter?" https://cosine.yale.edu/about-us/what-dark-matter.

4 *Britannica,* "dark matter," https://www.britannica.com/science/dark-matter. For instance, the article states: "Originally known as the 'missing mass,' dark matter's existence was first inferred by Swiss American astronomer Fritz Zwicky, who in 1933 discovered that the mass of all the stars in the Coma cluster of galaxies provided only about 1 percent of the mass needed to keep the galaxies from escaping the cluster's gravitational pull."

5 Yale University, COSINE-100 Dark Matter Experiment, "What is Dark Matter?" https://cosine.yale.edu/about-us/what-dark-matter.

6 Scoles, "Will Scientists Ever Find a Theory of Everything?" https://www.scientificamerican.com/article/will-scientists-ever-find-a-theory-of-everything/.

7 NASA Science Space Place, "What Is Dark Matter?" https://spaceplace.nasa.gov/dark-matter/en/.

8 Tyson, *Astrophysics for People in a Hurry*, 79.

9 Witsit Gets It, "Relatively Speaking – Episode 12 – The Death of Modern Cosmology," https://www.youtube.com/live/8CEEv4ezJBM, 57:00–58:00. Also see IAI TV, https://iai.tv/video/the-dark-matter-myth-pavel-kroupa. Specifically, he references "cold dark matter particles." Furthermore, see Kroupa, "The Dark Matter Crisis: Falsification of the Current Standard Model of Cosmology," https://www.cambridge.org/core/services/aop-cambridge-core/content/view/E3DB55A34BB36EA723671D1F54905B30/S1323358000001417a.pdf/dark_matter_crisis_falsification_of_the_current_standard_model_of_cosmology.pdf.

10 Ibid., 59:00–1:01:30. Also see IAI TV, https://iai.tv/video/the-dark-matter-myth-pavel-kroupa.

11 Ibid., 59:00–1:01:30. Also see IAI TV, https://iai.tv/video/the-dark-matter-myth-pavel-kroupa.

12 Ibid., 1:01:30–1:03:00. Also see IAI TV, https://iai.tv/video/the-dark-matter-myth-pavel-kroupa.

13 Gohd, "What Is Dark Energy? Inside our accelerating, expanding Universe," https://science.nasa.gov/universe/the-universe-is-expanding-faster-these-days-and-dark-energy-is-responsible-so-what-is-dark-energy/.

14 Ibid.
15 Albrecht et al., "Report of the Dark Energy Task Force," https://arxiv.org/pdf/astro-ph/0609591, 1.
16 Scoles, "Will Scientists Ever Find a Theory of Everything?" https://www.scientificamerican.com/article/will-scientists-ever-find-a-theory-of-everything/.
17 Ibid.
18 Robert Sungenis, "The Principle," https://www.youtube.com/watch?v=CeJb0JIHNik, 47:45–48:30.
19 Everyday Sounds, "Ken Campbell – Reality on the Rocks – Part 3 – Beyond our Ken," https://www.youtube.com/watch?v=6Qf0z5ZMjJs, 50:00-50:30.
20 For example, see Austin Whitsitt's content on his many social media channels, including his YouTube channel titled Witsit Gets It: https://www.youtube.com/@Whitsitt.
21 Ibid.
22 I first heard this comparison to Vedic philosophy made by Thomas Cowan, MD.
23 Nassim Nicholas Taleb popularized the term *black swan* with his book *The Black Swan: The Impact of the Highly Improbable* (second edition, 2010).
24 Andrew Kaufman, MD, "Fallacies, Fraud & Pseudoscience: The Parallels Between Cosmology and Biology," https://rumble.com/v44kwso-fallacies-fraud-and-pseudoscience-the-parallels-between-cosmology-and-biolo.html. 3:00–5:20.
25 Hawking and Mlodinow, *The Grand Design*, 56.
26 Witsit Gets It, "In The Field – Robert Bennett Ph.D. – Part 1," https://www.youtube.com/live/5j0wCrkzfwk, 12:45–15:10.
27 Figure reconstructed from NASA introduction to electromagnetic spectrum, https://science.nasa.gov/ems/01_intro.
28 Attenuation of light is a well-known phenomenon. For instance, "Fog attenuation in the visible and IR regions is reviewed from an empirical and theoretical point of view" is mentioned in Naboulsi et al., "Fog attenuation prediction for optical and infrared waves," https://proceedings.spiedigitallibrary.org/journals/optical-engineering/volume-43/issue-2/0000/Fog-attenuation-prediction-for-optical-and-infrared-waves/10.1117/1.1637611.short.
29 Clarke, *Profiles of the Future: An Inquiry into the Limits of the Possible*, 7.
30 Tesla, "Radio Power Will Revolutionize the World," https://teslauniverse.com/nikola-tesla/articles/radio-power-will-revolutionize-world.
31 A phrase used by Donald Rumsfeld. See Graham, "Rumsfeld's Knowns and Unknowns: The Intellectual History of a Quip," https://www.theatlantic.com/politics/archive/2014/03/rumsfelds-knowns-and-unknowns-the-intellectual-history-of-a-quip/359719/.

CHAPTER 1: SUN-CENTERED HISTORY AND PHILOSOPHY

1 Hoyle, *Nicolaus Copernicus: An Essay on his Life and Work*, 1.
2 Cohen, *The Birth of a New Physics*, 78, https://archive.org/details/B-001-015-439/page/n95/mode/2up?q=%22prove+that+the+earth%22.
3 For instance, see Mark, "Greek Astronomy," https://www.worldhistory.org/Greek_Astronomy/: "Greek astronomers rejected the heliocentric model of the universe, proposed by Aristarchus of Samos, because it contradicted the popular belief that the earth was the center of the universe."
4 Rieke (University of Arizona), "Ptolemy," https://ircamera.as.arizona.edu/NatSci102/NatSci102/lectures/ptolemy.htm.
5 *Britannica,* "Aristarchus of Samos," https://www.britannica.com/biography/Aristarchus-of-Samos.
6 *Britannica,* "Ptolemaic system," https://www.britannica.com/science/Ptolemaic-system.
7 Rovelli, *Anaximander*, 121.
8 For instance, see Kuhn, *The Copernican Revolution*; and Kuhn, *The Structure of Scientific Revolutions.*
9 *Stanford Encyclopedia of Philosophy,* "Nicolaus Copernicus," https://plato.stanford.edu/entries/copernicus/.
10 Wolf (UCLA), "The truth about Galileo and his conflict with the Catholic Church," https://newsroom.ucla.edu/releases/the-truth-about-galileo-and-his-conflict-with-the-catholic-church.
11 Ibid.
12 Ibid. Also see Moy, "Science, Religion, and the Galileo Affair," https://cdn.centerforinquiry.org/wp-content/uploads/sites/29/2001/09/22164801/p43.pdf: "After the trial, Galileo returned to his villa outside Florence, where he technically spent the last decade of his life under a very comfortable house arrest."
13 As cited in Question Everything, "Heliosorcery," https://rumble.com/v4ybjcq-heliosorcery-2022-exposing-the-occult-origins-of-heliocentrism-full-documen.html, 46:25. "Interview with Brother Guy Consolmagno" in *Astrobiology Magazine*. Available here, as well: https://www.economicsvoodoo.com/wp-content/uploads/2004-05-12-Interview-with-Brother-Guy-Consolmagno-Astrobiology-Magazine.pdf.
14 Wolf (UCLA), "The truth about Galileo and his conflict with the Catholic Church," https://newsroom.ucla.edu/releases/the-truth-about-galileo-and-his-conflict-with-the-catholic-church.
15 Sungenis, *Geocentrism 101*, 7–9. *Geocentrism 101* is an excellent resource for those interested in learning more about alternative takes on the history of astronomy and heliocentrism. Many of the ideas discussed briefly in chapters 1, 2, and 3 of this book are covered in more depth in *Geocentrism 101.*
16 Uri, "410 Years Ago: Galileo Discovers Jupiter's Moons," https://www.nasa.gov/history/410-years-ago-galileo-discovers-jupiters-moons/.

NASA Science Editorial Team, "Galileo's Observations of the Moon, Jupiter, Venus and the Sun," https://science.nasa.gov/solar-system/galileos-observations-of-the-moon-jupiter-venus-and-the-sun/.

17 An analogy given by many who question mainstream cosmology, such as David Weiss.

18 Sungenis, *Geocentrism 101*, 7.

19 For more on Tycho Brahe and his cosmology, see *Britannica*, "Tychonic system," https://www.britannica.com/biography/Tycho-Brahe-Danish-astronomer.

20 See Sungenis, *Geocentrism 101*, 26. More specifically, Kepler assumed elliptical orbits for the planets, rather than the circular ones used by Copernicus, which allowed for the nonuniform motion of planets—just like Ptolemy's geocentric model had proposed many centuries earlier.

21 Gilder and Gilder, *Heavenly Intrigue*, 3.

22 Ibid., 1.

23 BBC, "Astronomer Tycho Brahe 'not poisoned,' says expert," https://www.bbc.com/news/science-environment-20344201.

24 Gilder and Gilder, *Heavenly Intrigue*, 1.

25 For example, see NASA Earth Observatory, "Planetary Motion: The History of an Idea That Launched the Scientific Revolution," https://earthobservatory.nasa.gov/features/OrbitsHistory/.

26 Weinberg, *To Explain the World: The Discovery of Modern Science*, 251–52, https://archive.org/details/toexplainworlddi0000wein/page/252/mode/2up?q=proposition+43.

27 Assis, *Relational Mechanics*, 191.

28 As cited in Sungenis, *Geocentrism 101*, 43–44. Popov, "Newtonian–Machian analysis of the neo-Tychonian model of planetary motions."

29 Hoyle, *Nicolaus Copernicus: An Essay on His Life and Work*, 82.

30 Assis, *Relational Mechanics*, 261.

31 Einstein and Infeld, *The Evolution of Physics: From Early Concepts to Relativity and Quanta*, 224, https://archive.org/details/evolutionofphysi033254mbp.

32 Question Everything, "Heliosorcery," https://rumble.com/v4ybjcq-heliosorcery-2022-exposing-the-occult-origins-of-heliocentrism-full-documen.html.

33 Ibid., 47:25.

34 Robert Sungenis, *The Principle*, https://www.youtube.com/watch?v=CeJb0JIHNik.

35 Jonathan I. Katz, *The Biggest Bangs: The Mystery of Gamma-Ray Bursts, The Most Violent Explosions in the Universe*, 82.

36 Hawking, *A Brief History of Time*, 42. Robert Sungenis gives further context for this quote on p. 188 of *Geocentrism 101*: "The homogeneity that was needed to make Einstein and Hubble's universe feasible was calculated by a Russian physicist named Alexander Friedmann who adjusted Einstein's equations for this very purpose. A few decades later, Stephen Hawking admitted the real motivation behind Friedmann's

attempt to make the universe homogeneous. As was becoming common in modern cosmology, the motivation was to eliminate a center, and more specifically, Earth's possible occupation of the center." Hawking writes in *A Brief History of Time*: "There is, however, an alternate explanation [to a central Earth]: the universe might look the same in every direction as seen from any other galaxy, too. This, as we have seen, was Friedmann's second assumption. We have no scientific evidence for, or against, this assumption. We believe it only on grounds of modesty: it would be most remarkable if the universe looked the same in every direction around us, but not around other points in the universe."

37 As cited in Sungenis, *Geocentrism 101*, 30. The quotation is from a *Scientific American* article, and Sungenis includes a snapshot of the quotation from the article in his book.

38 Tyson, *Astrophysics for People in a Hurry*, 29–30.

39 Sungenis, "The Principle (2014 – Full Movie)," https://www.youtube.com/watch?v=CeJb0JIHNik, 1:14:00.

CHAPTER 2: HOW THE COSMOS BEGAN

1 For example, see NASA, "Edwin Hubble," https://science.nasa.gov/people/edwin-hubble/. Hubble also stated, "The assumption of motion, on the other hand, led to a non-linear law of red-shifts, according to which the velocities of recession accelerate with distance or with time counted backward into the past. A universe that has been expanding in this manner would be so extraordinarily young, the time-interval since the expansion began would be so brief, that suspicions are at once aroused concerning either the interpretation of red-shifts as velocity-shifts or the cosmological theory in its present form." Assis et al., "Hubble's Cosmology: From a Finite Expanding Universe to a Static Endless Universe," https://arxiv.org/pdf/0806.4481.

2 Hawking, *A Brief History of Time*, 39.

3 As summarized in Sungenis, *Geocentrism 101*, 180.

4 Hawking, *A Brief History of Time*, 42.

5 Hubble, *The Observational Approach to Cosmology*, 50–51. Also see Sungenis, *Geocentrism 101*, 179.

6 Ibid., 54.

7 Ibid., 58–59.

8 Sungenis, *Geocentrism 101*, 185.

9 For instance, see Sungenis, *Geocentrism 101*, 191.

10 Horgan, "Remembering Big Bang Basher Fred Hoyle," https://www.scientificamerican.com/blog/cross-check/remembering-big-bang-basher-fred-hoyle/.

11 Ibid.

12 NASA Hubblesite, "NASA's Webb, Hubble Telescopes Affirm Universe's Expansion Rate, Puzzle Persists," https://hubblesite.org/contents/news-releases/2024/news-2024-108.html.

13 Covered in Witsit Gets It, "Relatively Speaking – Episode 8 – Hubble Crisis," https://www.youtube.com/watch?v=wSe_GBITdwk, 8:00-9:30, 1:09:00–1:18:00.
14 NASA, "What Is the Inflation Theory?" https://wmap.gsfc.nasa.gov/universe/bb_cosmo_infl.html.
15 McRae, "Physicists Find No Sign of the Particle That Made the Universe Explode," https://www.sciencealert.com/physicists-find-no-sign-of-the-particle-that-made-the-universe-explode.
16 NASA, "Tests of Big Bang: The CMB," https://map.gsfc.nasa.gov/universe/bb_tests_cmb.html.
17 As cited in Sungenis, *Geocentrism 101*, 247–49.
18 As cited in Ibid.
19 Ibid., 248.
20 Edge, "THE ENERGY OF EMPTY SPACE THAT ISN'T ZERO," https://www.edge.org/conversation/the-energy-of-empty-space-that-isn-39t-zero.
21 Schwarz et al., "Is the Low-ℓ Microwave Background Cosmic?" in *Physical Review Letters*.
22 As cited in Sungenis, *Geocentrism 101*, 201.
23 As cited in Ibid., 203.
24 Ibid., 208.
25 Merali, "'Axis of evil' a cause for cosmic concern," https://www.newscientist.com/article/mg19425994-000 axis-of-evil-a-cause-for-cosmic-concern/.
26 Ibid.
27 Varshni, "The Red Shift Hypothesis for Quasars: Is the Earth the Center of the Universe?"
28 Sungenis, *Geocentrism 101*, 236.
29 Javanmardi and Kroupa, "Anisotropy in the All-Sky Distribution of Galaxy Morphological Types," https://arxiv.org/abs/1609.06719.
30 As cited in *Geocentrism 101*, 239–40.
31 Witsit Gets It, "In The Field – Robert Bennett Ph.D. – Part 1," https://www.youtube.com/live/5j0wCrkzfwk, 2:41:00–2:43:00.

CHAPTER 3: EARTH'S MOTION AND THE AETHER

1 Witsit Gets It, "In The Field – Robert Bennett Ph.D. – Part 1," https://www.youtube.com/live/5j0wCrkzfwk, 2:43:00.
2 Michels, "Granite and Grace at Llano's World Championship of Rock Stacking," https://www.texasmonthly.com/being-texan/llano-world-rock-stacking-championship/.
3 For example, see *National Geographic*, "Rotation," https://education.nationalgeographic.org/resource/rotation/; Na et al., "Chandler Wobble and Free Core Nutation: Theory and Features," https://adsabs.harvard.edu/pdf/2019JASS...36...11N#:~:text=Chandler%20wobble%20is%20a%20prograde,latter%20is%20free%20core%20nutation; Howell and Urrutia, "How fast is Earth moving?" https://www.space.com/33527-

how-fast-is-earth-moving.html; McFall-Johnsen, "Earth is screaming through space at 1.3 million mph. A simple animation by a former NASA scientist shows what that looks like," https://www.businessinsider.com/earth-screaming-through-space-nasa-animated-video-2019-10.

4 Rosenberg, "Speed of the Earth," https://www.thoughtco.com/speed-of-the-earth-1435093.

5 *Level Earth Observer*, "Flat Earth: I love destroying the globe." https://www.youtube.com/watch?v=R5PPaJkmpDc.

6 Snow, "New World Record Set for Farthest Long-Range Rifle Shot: 4.4 Miles," https://www.outdoorlife.com/guns/new-world-record-longest-rifle-shot/.

7 jeranism, "Long Range Shooting World Record Broken...What Was The Coriolis Correction?." https://www.youtube.com/watch?v=xPkCtAvNW2o.

8 Casto, "Which Direction Does Toilet Water Swirl at the Equator?," https://www.livescience.com/33567-toilet-swirl-direction-equator.html.

9 Einstein, "How I created the theory of relativity," https://courses.physics.ucsd.edu/2022/Fall/physics2d/einsteinonrelativity.pdf.

10 Ault, "How Does Foucault's Pendulum Prove the Earth Rotates?" https://www.smithsonianmag.com/smithsonian-institution/how-does-foucaults-pendulum-prove-earth-rotates-180968024/.

11 Sungenis, *Geocentrism 101*, 72–74.

12 Einstein, "To Ernst Mach," https://einsteinpapers.press.princeton.edu/vol5-trans/362. Einstein wrote: "If one accelerates an inertial spherical shell S, then, according to the theory, a body enclosed by it experiences an accelerating force.…If the shell S rotates about an axis passing through its center (relative to the fixed stars ('Restsystem'), then a Coriolis field arises inside the shell, i.e., the plane of the Foucault pendulum is being carried along (though with a practically immeasurably small velocity)."

13 Assis, *Relational Mechanics*, 223.

14 For instance, see Priest and Pechan, "The driving mechanism for a Foucault pendulum (revisited)," https://pubs.aip.org/aapt/ajp/article-abstract/76/2/188/1057096/The-driving-mechanism-for-a-Foucault-pendulum?redirectedFrom=fulltext. *Smithsonian Magazine* also refers to an electromagnetic push: "Looking down, visitors would see a symmetrical hollow brass bob weighing about 240 pounds and shaped like an inverted teardrop. As it moved back and forth—facilitated by an electromagnetic push to keep it continuously swinging despite air resistance and vibrations in the cable—it would knock down inch-or-so-high pins standing at fixed points along the circumference of a small circle. Over time, viewers could see the direction of the pendulum's swing change, implying that the Earth was rotating beneath them." See Ault, "How Does Foucault's Pendulum Prove the Earth

Rotates?" https://www.smithsonianmag.com/smithsonian-institution/how-does-foucaults-pendulum-prove-earth-rotates-180968024/.
15 Mitchell fromAustralia, "At Best, All YOU'VE Done, Is Debunk The Rotation of the Earth!" https://www.youtube.com/watch?v=sQNWMuqqsFQ.
16 Witsit Gets It, "In The Field – Robert Bennett Ph.D. – Part 1," https://www.youtube.com/live/5j0wCrkzfwk, 47:30–48:30.
17 Ibid., 50:00–50:30.
18 FlatEarth.ws, "Negative Parallax," https://flatearth.ws/negative-parallax.
19 See Sungenis, *Geocentrism 101*, 67, for more on stellar aberration.
20 Henri Poincaré, "The Principles of Mathematical Physics," *The Monist*, vol. XV, January 1905, 20; http://www.jstor.org/stable/27899559?seq=20.
21 *Britannica*, "ether," https://www.britannica.com/science/ether-theoretical-substance.
22 Nedic, "Thermo-Gravitational Oscillator and Tesla's Energy Source Ether," https://www.researchgate.net/publication/280304207_Thermo-Gravitational_Oscillator_and_Tesla's_Energy_Source_Ether.
23 As one example, see Davidson, "Free Energy, Gravity and the Aether," http://tuks.nl/pdf/Reference_Material/Paul_Stowe/Free%20Energy,%20Gravity%20and%20the%20Aether%20101897.pdf. Also see *Occult Ether Physics: Tesla's "Ideal Flying Machine" and the Conspiracy to Conceal It* by William Lyne.
24 As cited, with images of the document itself, in Odle, *Like Clay Under the Seal*, 378–79.
25 Witsit Gets It, "In The Field – Robert Bennett Ph.D. – Part 1," https://www.youtube.com/live/5j0wCrkzfwk, 1:00:00-1:03:00.
26 Ibid., 1:07:30–1:09:10.
27 Lorentz, "On the Influence of the Earth's Motion on Luminiferous Phenomena" in Miller, *Albert Einstein's Special Theory of Relativity*, 20.
28 Witsit Gets It, "In The Field – Robert Bennett Ph.D. – Part 1," https://www.youtube.com/live/5j0wCrkzfwk, 1:52:00-1:53:00.
29 Michelson, "The Relative Motion of the Earth and the Luminiferous Ether," 125.
30 For a discussion of the Michelson studies, see Witsit Gets It, "In The Field – Robert Bennett Ph.D. – Part 1," https://www.youtube.com/live/5j0wCrkzfwk, 1:53:00. Also see Sungenis, *Geocentrism 101*, chapter 5 ("The Michelson-Morley Experiment Confirms Earth Is Motionless").
31 As cited in Sungenis, *Geocentrism 101*, 103. *Science in History*: Volume 3, 744 (from Jaffe, 88).
32 Witsit Gets It, "In The Field – Robert Bennett Ph.D. – Part 1," https://www.youtube.com/live/5j0wCrkzfwk, 1:54:00–1:57:00.
33 Ibid., 1:58:00–1:59:00. Whitsitt references many replications of the fringe shift as well. See his presentation shown at Kim Osbøl – Copenhagen Denmark, "Anti-Disinformation: True Earth 101 – Motion! [23.11.2023]," https://www.bitchute.com/video/ljdtJOl9Z3Ix/, 8:34.

34 *Britannica*, "Michelson-Morley experiment," https://www.britannica.com/science/Michelson-Morley-experiment.
35 From Poincaré's report *La science et l'hypothèse* ("Science and Hypothesis") 1901, 1968, p. 182. L. Kostro's, Einstein and the Ether, 2000, p. 30.
36 Henri Poincaré, "The Principles of Mathematical Physics," *The Monist*, vol. XV, January 1905, 6-7; http://www.jstor.org/stable/27899559?seq=20.
37 G. J. Whitrow, *The Structure and Evolution of the Universe*, 1949, 1959, 79.
38 A screenshot of the quotation from the book is shown in Sungenis, *Geocentrism 101*, 102. It can be found in Pauli, *Theory of Relativity*, 4.
39 Jaffe, *Michelson and the Speed of Light*, 76.
40 James A. Coleman, *Relativity for the Layman*, p. 37. Sungenis remarks on page 279 of *Geocentrism 101*: "Of Coleman's book Einstein wrote: "Gives a really clear idea of relativity" (front cover of the 1954 edition).
41 Baker, *Modern Physics & Antiphysics*, 53–54.
42 As cited in Sungenis, *Geocentrism 101*, 108.
43 Witsit Gets It, "In The Field – Robert Bennett Ph.D. – Part 1," https://www.youtube.com/live/5j0wCrkzfwk, 2:03:00–2:04:00.
44 Ibid., 2:17:00–2:21:00.
45 Einstein, "How I created the theory of relativity," https://courses.physics.ucsd.edu/2022/Fall/physics2d/einsteinonrelativity.pdf. "This problem" refers to the motion of Earth as shown in the paragraph preceding the quoted material in the text of this book. Einstein's paragraph states: "Then I myself wanted to verify the flow of the ether with respect to the Earth, **in other words, the motion of the Earth**. When I first thought about **this problem**, I did not doubt the existence of the ether or the motion of the Earth through it. I thought of the following experiment using two thermocouples: Set up mirrors so that the light from a single source is to be reflected in two different directions, one parallel to the motion of the Earth and the other antiparallel. If we assume that there is an energy difference between the two reflected beams, we can measure the difference in the generated heat using two thermocouples. Although the idea of this experiment is very similar to that of Michelson, I did not put this experiment to the test." [bold added here in order to clarify the use of the words "this problem" in the bracketed text of this book]
46 A screenshot from Ronald Clark's book is provided in Sungenis, *Geocentrism 101*, 147.
47 Einstein, "How I created the theory of relativity," https://courses.physics.ucsd.edu/2022/Fall/physics2d/einsteinonrelativity.pdf.
48 For instance, see Sungenis, *Geocentrism 101*, 113–14, 119.
49 Skeptics of heliocentrism have remarked along these lines, including Austin Whitsitt. Also see Sungenis, *Geocentrism 101*, chapter 6.
50 For more on relativity, see *Britannica*, "relativity," https://www.britannica.com/science/relativity.

51 Sungenis, *Geocentrism 101*, 142.

52 Poor, *Gravitation versus Relativity*, 261. A screenshot of the quotation from the book is also provided at Sungenis, *Geocentrism 101*, 148. It's worth noting that Einstein admitted to the reality of the aether. He wrote in a 1919 letter to Lorentz: "It would have been more correct if I had limited myself, in my earlier publications, to emphasizing only the non-existence of an aether velocity, instead of arguing the total non-existence of the aether, for I can see that with the word aether we say nothing else than that space has to be viewed as a carrier of physical qualities." And in 1920, he stated in a lecture: "Recapitulating, we may say that according to the general theory of relativity, space is endowed with physical qualities; in this sense, therefore, there exists an ether. According to the general theory of relativity, space without ether is unthinkable." However, the caveat for Einstein's conception of the aether was that "the idea of motion may not be applied to it." See Einstein, as cited in Rafelski, "Critical Acceleration and Vacuum Structure," PDF p. 16, http://www.physics.arizona.edu/~rafelski/PS/1112Whitten UCLA.pdf. and https://www.spaceandmotion.com/Physics-Albert-Einstein-Leiden-1920.htm. Also see Sungenis, *Geocentrism 101*, 162–63.

53 See Sungenis, *Geocentrism 101*, chapter 6 ("Einstein's Special Relativity: Invented to Keep Earth Moving"). Sungenis also notes on p. 140: "No one has ever proven that light travels the same speed in deep space as it does on the surface of the Earth."

54 Einstein, *Relativity: The Special and General Theory*, 89, https://archive.org/details/relativitythespe30155gut.

55 Witsit Gets It, "In The Field – Robert Bennett Ph.D. – Part 1," https://www.youtube.com/live/5j0wCrkzfwk, 2:38:00–2:39:00. In summary, Bennett notes that Einstein's special and general relativity theories have different restrictions in two critical ways: General relativity allows for a form of aether (without motion), whereas special relativity doesn't; and general relativity doesn't have a truly constant speed of light, whereas special relativity does. Special relativity is supposed to be a special case within general relativity, so the mismatch is highly problematic.

56 Ibid., 7:00–10:00.

57 For instance, see Weiss, "Mercury's Orbital Precession, General Relativity, and the Solar Bulge," https://math.ucr.edu/home/baez/physics/Relativity/GR/mercury_orbit.html.

58 Kroupa and Haslbauer, "Our model of the universe has been falsified," https://iai.tv/articles/our-model-of-the-universe-has-been-falsified-auid-2393. The article states: "A recent publication in the journal Physical Review D with about 156 co-authors suggests the distribution of matter to be smoother than expected, based on the predictions of the standard model of cosmology. This new data release by the Dark Energy Survey, was based on the findings of a telescope in Chile that measured the tiny distortions of the images of relatively nearby galaxies,

caused by their light being diverted due to the gravitational pull of foreground matter."

59 Young, *A Fool's Wisdom*, 62.

60 Ibid., 61.

61 Richard Feynman lecture, *Electricity in the Atmosphere*, https://www.feynmanlectures.caltech.edu/II_09.html.

62 See https://www.nasa.gov/wp-content/uploads/2016/01/g-29859_esl.pdf and https://ntrs.nasa.gov/citations/20150022349.

63 Electrostatics mentioned in an article by Hutchinson in Electric Spacecraft Journal (p. 21), https://www.thehutchisoneffect.com/docs/Hutchison%20Effect%20-%20More%20Info.pdf: "Figure 17 is another pulsing circuit devices I call the interferometer. Electrostatics were produced with two Van de Graaff generators. When arranged as shown in Figure 18, cool wind effects could be produced. These machines were used to impose an electrostatic field in the test area."

64 Wood vs. Applied Research Associates, Inc. Affidavit, Hutchinson. https://www.thehutchisoneffect.com/docs/AffJHutchison4.pdf.

65 Witsit Gets It, "Schooling Globers – Episode 25," https://www.youtube.com/live/3yYdQbKCSC8, 34:00–54:00. Also see many studies on thunderstorm anomalies posted at the Aether Cosmology website, https://aethercosmology.com/t/downward-acceleration-gravity-anomalies-thunderstorms/134.

66 Recall Robert Bennett's quotation mentioned in the introduction: "When we design a test, the test has to match what the hypothesis claims. So, if you claim something that is going to apply throughout all space, you cannot test it only on Earth. If you say it's in the Solar System, you have to be able to test it throughout the Solar System. This is what's overlooked very often in the statements of laws—the scope of the laws—in space." Also, the example of "spheres rotating around spheres" is one that has been mentioned by many others, and here I am merely summarizing this commonly referenced point.

67 I am summarizing here an argument made by many others.

68 These are examples used by many advocates of alternative cosmologies, including Eric Dubay (among others).

69 Virginia Institute of Marine Science, "The Equilibrium Theory of Tides," https://www.vims.edu/research/units/labgroups/tc_tutorial/static.php.

70 Young, *A Fool's Wisdom*, 61.

CHAPTER 4: SKEPTICISM ABOUT THE NATURE OF OUR HOME

1 NASA's Earth Observatory, "The Blue Marble," https://earthobservatory.nasa.gov/features/BlueMarble/BlueMarble_2002.php.

2 The 92nd Street Y, New York, "Neil deGrasse Tyson explains how the earth became pear-shaped," https://www.youtube.com/watch?v=SoCKapivHGM.

3 Note that some geocentrists, such as Ptolemy, believed in a spherical Earth.

4 Sungenis, *Flat Earth/Flat Wrong*, 327.
5 Young, *A Fool's Wisdom*, 51–53, 54.
6 See the works of Austin Whitsitt, David Weiss, and Eric Dubay for more information on many of these ideas. There are, of course, many other researchers, but the three mentioned here have particularly comprehensive bodies of work and have amassed large followings. David Weiss's Flat Earth Sun, Moon & Zodiac Clock App is a helpful starting point. See https://www.flatearth101.com/fe-david-weiss, for example. The bibliography contains many references as well. Although there might be slight differences of opinion on various matters, the overarching framework is generally endorsed by many Globe Skeptics.
7 As cited in Hamer, *The Falsification of Science*, 175.
8 Taboo Conspiracy, "How Time Zones Hide the Flat Earth by Hervé Riboni," https://www.youtube.com/watch?v=lMhheDWThxE, 6:30–7:30.
9 Some speculate that a giant magnet at the North Pole generates Earth's magnetic field. However, the exact mechanism by which the magnetic field is generated in a Flat Earth framework requires further investigation of Earth.
10 Rovelli, *Anaximander*, 154.
11 Credit to Austin Whitsitt for making this point clear so often.
12 As noted by Lee, *Flat Earth*, 14.
13 Witsit Gets It, "Schooling Globers – Episode 33 (Geocentrism)," https://www.youtube.com/live/oY1svV3dMVg, 33:00.
14 Young, *A Fool's Wisdom*, 62.
15 Again, these ideas are promoted by many Globe Skeptics. Austin Whitsitt, David Weiss, and Eric Dubay are examples, among many others.
16 Ibid.
17 CIA-RPD86-00513R00134372008-3, https://www.cia.gov/readingroom/docs/CIA-RDP86-00513R001343720008-3.pdf, 19–20.
18 As cited in Odle, *Like Clay Under the Seal*, 207–08.
19 Online Etymology Dictionary, "planet (n.)," https://www.etymonline.com/word/planet.
20 Flat Earth Dave Interviews 2, "SEASONS EXPLAINED on a Flat Earth – Justin Peters Ministries." https://www.youtube.com/watch?si=GD_qdr5flZyanXwL&v=ggzJnG9ERjM&feature=youtu.be. Also see Bikos and Kher, "Twilight, Dawn, and Dusk," https://www.timeanddate.com/astronomy/different-types-twilight.html.
21 Witsit Gets It, "'Professor Dave' vs. Witsit–Flat Earth Debate Review," https://www.youtube.com/watch?v=qc1oG6yzJYg&t=11297s, 1:05:39.
22 For instance, government video feeds of Antarctica are discussed and shown here in Flat Earth Dave Interviews, "Brad Olson discusses Flat Earth but remains lost in Antarctica," https://youtu.be/IsWJWvSKrac?si=jCs7AU7DU9QBMSgV, 1:49:00–1:51:00; 1:55:00; 1:59:45.

Also see Paul On The Plane, "Antarctica is NOT a continent | Tiger Dan925 Jumped Ship (Jeranism Mirror)," https://www.youtube.com/watch?v=YLkTm19aZGs, 21:30-30:00; and Flat Earth Sun, Moon, & Zodiac Clock app, "Antarctica 24 Hour Sun FAKED again!," https://www.youtube.com/watch?v=o7uLeVKv0Sc. Austin Whitsitt describes the position here when he says the two most famous "twenty-four-hour Antarctic Summer Sun" videos were "faked" so skeptics know that it's possible. See Witsit Gets It, "Schooling Globers – Episode 35 (Logic)" https://www.youtube.com/live/9cRWrHuPYp8, 3:38:00-3:40:00.

23 Witsit Gets It, "Schooling Globers – Episode 7 – Azimuthal Grid of Vision," https://www.youtube.com/watch?v=HV89DJhCsCE&t=5672s, 51:00–52:00.

24 Ibid.

25 Ibid., 43:00–50:00.

26 Witsit Gets It, "Professor Dave" vs. Whitsitt – Flat Earth Debate Review," https://www.youtube.com/watch?v=qc1oG6yzJYg&t=2995s, 50:00–51:00.

27 Witsit Gets It, "Schooling Globers – Episode 7 – Azimuthal Grid of Vision," https://www.youtube.com/live/HV89DJhCsCE, 2:07:00–2:10:00.

28 In this model, everything is interconnected and electric or electrostatic, possibly compressed by the aether. Young notes: "In the heliocentric globe model, the spinning planet is deemed to be the generator of the toroidal vortex…but in the geocentric perspective, life manifests from the vortex, on the plane of inertia [that is, what we call 'Earth'], which is the level surface in the centre of the torus." Young, *A Fool's Wisdom*, 253.

29 Young, *A Fool's Wisdom*, 80.

30 Mentioned by Austin Whitsitt on an "X Live" event hosted by Braveritas on June 4, 2024; also available at https://www.youtube.com/watch?v=ssCZXJrDxiQ.

31 Celebrate Truth, "Rob Skiba vs. Dr. Robert Sungenis | Flat Earth Biblical Debate 2018," https://www.youtube.com/watch?v=AqbiwtRKrtg.

32 Discussed in detail at Witsit Gets It, "Schooling Globers – Episode 34 (The Core of the Globe Belief)," https://www.youtube.com/watch?v=lTIFsHH5i6U&t=2708s.

33 Leeman, "The Deepest Borehole on Earth – Don't Panic Podcast Episode 350," https://leemangeophysical.com/the-deepest-borehole-on-earth/.

34 Freedman, "What percentage of the ocean have we mapped?" https://www.livescience.com/planet-earth/rivers-oceans/what-percentage-of-the-ocean-have-we-mapped.

35 See, for example, RAND Corporation, "U.S. Armed Forces Capabilities in Arctic Region Pose National Security Risks, Need Strengthening," https://www.rand.org/news/press/2023/11/01.html.

36 For instance, see Nordregio, "Protected areas in the Arctic," https://nordregio.org/maps/protected-areas-in-the-arctic/. The article explicitly

states: "Within the northern circumpolar permafrost region, there are ca. 1300 protected areas." Also see the Arctic Council's website, which discusses "Protected areas in the Arctic region": https://arctic-council.org/news/protected-areas-in-the-arctic-region/.

37 See CBC News, "NORAD upgrades a 'good move' for northern security, says Nunavut MP," https://www.cbc.ca/news/canada/north/norad-upgrade-defence-nunavut-nws-1.6498070.

38 For example, see *Britannica*, "International Geophysical Year," https://www.britannica.com/event/International-Geophysical-Year.

39 Taboo Conspiracy, "Sorry, Antarctica Is Closed," https://www.youtube.com/watch?v=SmYRFtY_jfQ.

40 Witsit Gets It, "Schooling Globers – Episode 13 – Magnetic Declination," https://www.youtube.com/live/2BLIkqri2hg, 59:00–1:00:00.

41 "Report of the Signal Corps Observer: Operation 'Highjump,'" http://www.bahaistudies.net/asma/Operation_Highjump.pdf.

42 Project Petri, "Admiral Richard E. Byrd 1954 interview, enhanced audio," https://www.youtube.com/watch?v=PsheO3qh2dw.

43 Dunham, "4 Reasons Why You Can't You Fly Over Antarctica (and 4 Exceptions)," https://executiveflyers.com/why-cant-you-fly-over-antarctica/.

44 Morgan, *The Iron Republic*, 2.

45 Young, *A Fool's Wisdom*, 67.

46 For example, see Sokol, "How the Ancient Art of Eclipse Prediction Became an Exact Science," https://www.quantamagazine.org/how-the-ancient-art-of-eclipse-prediction-became-an-exact-science-20240405/.

47 Garwood, *Flat Earth*, 15–16.

48 Ibid., 16.

49 Ibid., 17.

50 Charan, *Vedic Universe: Flat Earth*, 21.

51 Hannam, "A Square, Flat Earth and Round Heaven Meant the World to Ancient China," https://www.atlasobscura.com/articles/flat-earth-ancient-china.

52 Garwood, *Flat Earth*, 19.

53 Hoskin, *The History of Astronomy: A Very Short Introduction*, 10.

54 Mike Boll, "Neil deGrasse Tyson Destroys The Eratosthenes Story," https://www.youtube.com/watch?v=Gs0M0FJbKI4.

55 Rovelli, *Anaximander*, 152.

56 For more on this topic, see David Weiss's app, under the Frequently Asked Questions box titled "Sticks and Shadows."

57 For more on the history of Flat Earth, see *Off the Edge* by Kelly Weill and *Flat Earth* by Chrstine Garwood.

58 Planet Plane, "Eric Dubay: The History of Flat Earth," https://www.youtube.com/watch?v=9496sa0cXh4, 1:19:00.

59 Ibid., 1:19:00–1:21:00.

60 Ibid., 1:22:20.

61 Ibid., 1:22:30.

62 Weill, *Off the Edge*, 90–91.
63 See, for example, Sungenis, *Flat Earth/Flat Wrong*, 57.
64 *Shoot the Moon* documentary by Crrow777, https://vimeo.com/ondemand/shootthemoon.
65 See https://trueearther.com/.
66 Witsit Gets It, "Flat Earth Debate Review: Whitsitt vs. Harrison," https://www.youtube.com/live/QDCzNoX1Uzg?si=oSZyfJUEqnRRRJT3.
67 Smith, "Travis And Jason Kelce Reveal How Many NFL Players Believe In 'Flat Earth' Conspiracy Theory: 'You Would Be Shocked,'" https://www.msn.com/en-us/sports/nfl/travis-and-jason-kelce-reveal-how-many-nfl-players-believe-in-flat-earth-conspiracy-theory-you-would-be-shocked/ar-BB1m2K92.
68 France24, "To 11 million Brazilians, the Earth is flat," https://www.france24.com/en/20200228-to-11-million-brazilians-the-earth-is-flat.
69 David Weiss's Flat Earth Sun, Moon & Zodiac Clock app, at https://theflatearthpodcast.com/, includes videos of such professionals, among many other basics related to Flat Earth concepts (in particular, see the Frequently Asked Questions playlist of videos under the heading "Pilots and Scientists?").
70 Witsit Gets It, "Schooling Globers – Episode 23," https://www.youtube.com/live/7QYaxbsVgd0, 2:13:00-2:14:00.
71 Ibid., 2:40:00.
72 Witsit Gets It, "NASA Fraud – Episode 3," https://www.youtube.com/live/_gpM5lmRVW8, 1:37:00–1:39:00. Whitsitt says on the show that this former NASA employee will come on his show. As of the writing of this book, that has not yet happened.
73 Spectator staff, "Is the Earth round or flat? Culture Ministry chief official says proof is still needed," https://spectator.sme.sk/c/23325477/is-the-earth-round-or-flat-culture-ministry-chief-official-says-proof-is-still-needed.html?ref=njctse.
74 Taboo Conspiracy, "USSR Cosmonauts: Earth Is Flat and Nobody Has Been to Space (Old Taboo)," https://www.youtube.com/watch?v=2oUzONrecXc.
75 Malashenko, "Fact Check: Polish Astronaut Did NOT Admit the Earth Is Flat — It Was a Joke," https://leadstories.com/hoax-alert/2023/07/fact-check-polish-astronaut-did-not-admit-the-earth-is-flat.html.
76 Presented at the end of *Flatlanders* documentary, Ep. 3.
77 June 17, 2014 conversation between Mark Gober and Google's Gemini AI.
78 Flat Earth Sun, Moon & Zodiac Clock app, "NETFLIX Debunks Flat Earth – Ring Laser Gyro and Laser Experiment deception," https://youtu.be/XHh2X_dWdxA?si=CuDGvdiGKjlJm-hL. Also explained by David Weiss in his 2024 interview with Roseanne Barr. See Roseanne Barr, "What's beyond Antarctica? with Flat Earth Dave | The Roseanne Barr Podcast #55," https://www.youtube.com/watch?v=bIeOPtHY4KI.

CHAPTER 5: GLOBE CHALLENGES

1 EndeavorFreedom, "Rob Skiba and Zen Garcia – FE Court Case and the Great Deception," https://www.youtube.com/watch?v=lv2-T3nKjb0, 16:00-17:00.
2 The legal document ("Case No.: 2019-MV-1104") is available at the following link: https://rodscontracts.com/docs/legal/Thompson VersusGarcia.pdf.
3 The distance cited by Garcia is based on the following: "The London and Northwestern Railway forms a straight line 180 miles long between London and Liverpool."
4 As cited in a legal document for Zen Garcia's case ("Case No.: 2019-MV-1104"), available at the following link: https://rodscontracts.com/docs/legal/ThompsonVersusGarcia.pdf.
5 As cited in Ibid.
6 As cited in Ibid.
7 Odle, *Like Clay Under the Seal*, 442–45. https://apps.dtic.mil/sti/tr/pdf/AD0020861.pdf.
8 James and Harris, "Calculation of Wind Compensation for Launching of Unguided Rockets," 7, https://ntrs.nasa.gov/api/citations/20040008097/downloads/20040008097.pdf.
9 Duke et al., "Derivation and Definition of a Linear Aircraft Model," 30. https://ntrs.nasa.gov/api/citations/19890005752/downloads/19890005752.pdf.
10 Miletta, "Propagation of Electromagnetic Fields Over Flat Earth." https://ia903206.us.archive.org/23/items/propagation_of_electromagnetic_fields_over_flat_earth/propagation_of_electromagnetic_fields_over_flat_earth.pdf
11 Cooper, "Adding Liquid Payloads Effects to the 6-DOF Trajectory of Spinning Projectiles," 1. https://apps.dtic.mil/sti/pdfs/ADA519118.pdf.
12 Tunick, "An Energy Budget Model to Calculate the Low Atmosphere Profiles of Effective Sound Speed at Night," 16, https://apps.dtic.mil/sti/pdfs/ADA414792.pdf.
13 Aerostudents.com, "Flight Dynamics Summary," https://www.aerostudents.com/courses/flight-dynamics/flightDynamicsFullVersion.pdf.
14 Witsit Gets It, "Schooling Globers – Episode 16 – Sonar," https://www.youtube.com/live/I-4Wp3e8kIg.
15 Coomes, "Mirage of Chicago skyline seen from Michigan shoreline," https://www.abc57.com/news/mirage-of-chicago-skyline-seen-from-michigan-shoreline.
16 As quoted in a legal document for Zen Garcia's case ("Case No.: 2019 MV-1104"), available at the following link: https://rodscontracts.com/docs/legal/ThompsonVersusGarcia.pdf. For more information on this case study, see Rob Skiba, "The Chicago Skyline Expanded Edition–Part 1: Refraction, Magnification or Curvature?" https://www.youtube.com/watch?v=I-Q-FuXJSTQ&t=1s.

17 For instance, see Hamer, *The Falsification of Science*, 198-204; and Hendrie, *The Greatest Lie on Earth*, 25–50.

18 Whitsitt X account, June 4, 2024, https://x.com/Witsitgetsit/status/1798165881429668322?s=46&t=ijTvU_628uOjAUiS9U8Hxw, 2:51:00. Also partially available at https://www.youtube.com/watch?v=ssCZXJrDxiQ (so the time stamps do not perfectly match).

19 Ibid., 2:52:00.

20 Ibid., 2:57:00–2:53:00.

21 Ibid., 2:54:00.

22 Ibid., 2:57:00–2:58:00.

23 Ibid., 2:57:00–2:59:00.

24 Anti-Disinfo Leauge [sic], "True Earth 101: Curvature," https://www.bitchute.com/video/6nbYsiGpwAH9/, 5:00-6:00.

25 RobinsHoodlum, "Globe Earth SNIPER ~ from Mountainbear," https://rumble.com/v1hlc50-globe-earth-sniper-from-mountainbear.html.

26 Garwood, *Flat Earth*, 20.

27 Mitchell fromAustralia, "Flat Earth School – Why Objects Disappear Bottom Up," https://www.youtube.com/watch?v=aVVbsekJ9Sg. Also see Witsit Gets It, "Schooling Globers – Episode 26," https://www.youtube.com/live/FBKrYJ0YvUY, 1:59:00-2:00:00.

28 Witsit Gets It, "In The Field – Robert Bennett Ph.D. – Part 1," https://www.youtube.com/live/5j0wCrkzfwk, 2:11:00-2:12:00.

29 Norling, *Perspective Made Easy*, 9.

30 Young, *A Fool's Wisdom*, 79–80.

31 Odle, *Like Clay Under the Seal*, 100. Odle includes the image from the magazine itself that includes the text. It's worth noting that Sungenis cites an alternative view in his book *Flat Earth/Flat Wrong*, p. 12. He cites a 2008 article in *Applied Optics* in which the author, David K. Lynch, writes: "The first direct visual detection of the curvature of the horizon has been widely attributed to Auguste Piccard and Paul Kipfer on 27 May 1931." Sungenis cites this as if it's contradictory evidence to the *Popular Science* quote about Piccard. Piccard saw "upturn" edges, not the downturn edges that would be expected on a Globe. So perhaps that is what Lynch was referring to. Otherwise, Lynch does not cite a primary source suggesting that Piccard saw downward curvature, beyond noting the following: "Piccard is widely believed to be the first. There are many references to his achievement on the Internet, most of them certainly derivative. I contacted the Piccard family and they were aware of the claim but had no hard evidence or literature citation backing it up." Thus, the claim that Piccard actually saw downward curvature suggestive of a Globe seems unsubstantiated.

32 Odle, *Like Clay Under the Seal*, 102. Odle includes an image from the primary source, as well.

33 Whitsitt X account, June 4, 2024, https://x.com/Witsitgetsit/status/1798165881429668322?s=46&t=ijTvU_628uOjAUiS9U8Hxw,

3:01:00–3:04:00. Also available at https://www.youtube.com/watch?v=ssCZXJrDxiQ.

34 Witsit Gets It, "Schooling Globers – Episode 16 – Sonar," https://www.youtube.com/live/I-4Wp3e8kIg. Also see Flat Earth Sun Earth Moon & Zodiac Clock App, "Humpback Whales Prove Our Flat Earth," https://www.youtube.com/watch?v=KIjO1_KCXJA. Also see American Oceans, "How Do Whales Communicate?," https://www.americanoceans.org/facts/how-do-whales-communicate/.

35 Aether Cosmology, "Long-Distance Specular Reflections Illustrate a Lack of Curvature," https://aethercosmology.com/t/long-distance-specular-reflections-illustrate-a-lack-of-curvature/95.

36 Ibid.

37 Flat Earth, Banjo, USA, Japan, and Brazil, "Falling Stars Prove FLAT EARTH," https://www.youtube.com/watch?v=Ei8lfNHus1M. Also see, DITRH, "Meteors determine up and down," https://www.youtube.com/watch?v=RhQHV0AAFrU.

38 Flat Earth Sun Moon & Zodiac Clock App, "Meteorite crater or guizer?", https://www.youtube.com/watch?si=nY5vjwFe_fMcyeAY&v=Tq7MFz_aQog&feature=youtu.be. Flat Earth Sun Moon & Zodiac Clock App, "METEORS AND METEOR SHOWERS," https://www.youtube.com/watch?si=4EMy1qbZliSGxCoB&v=Ir-8rptN-k8&feature=youtu.be.

39 *NOVA* PBS Official, "Arctic Sinkholes I Full Documentary I NOVA I PBS," https://www.youtube.com/watch?v=HvKpnaXYUPU, 3:00-4:00.

40 Whitsitt X account, June 4, 2024, https://x.com/Witsitgetsit/status/1798165881429668322?s=46&t=ijTvU_628uOjAUiS9U8Hxw, 3:53:00–3:54:00.

41 Flat Earth Sun Moon & Zodiac Clock App, "METEORS AND METEOR SHOWERS," https://www.youtube.com/watch?v=Ir-8rptN-k8.

42 For more on this topic, see Witsit Gets It, "Schooling Globers – Episode 34 (The Core of the Globe Belief)," https://www.youtube.com/watch?v=lTIFsHH5i6U&t=10005s.

43 University of Maryland Department of Physics, "Lathrop Lab's Geodynamo Set for Overhaul," https://umdphysics.umd.edu/about-us/news/research-news/1562-lathrop-lab-s-geodynamo-set-for-overhaul.html.

44 Humphreys, "Planetary Magnetic Dynamo Theories: A Century of Failure," https://digitalcommons.cedarville.edu/cgi/viewcontent.cgi?article=1212&context=icc_proceedings.

45 Witsit Gets It, "Schooling Globers – Episode 34 (The Core of the Globe Belief)," https://www.youtube.com/watch?v=lTIFsHH5i6U&t=10005s, 39:00–41:00.

46 Some speculate that a giant magnet at the North Pole generates Earth's magnetic field. However, the exact mechanism by which the magnetic field is generated in a Flat Earth framework requires further investigation of Earth.

47 For a detailed discussion of magnetic declination, see Witsit Gets It, "Schooling Globers – Episode 13 – Magnetic Declination" https://www.youtube.com/live/2BLIkqri2hg.
48 Witsit Gets It, "Schooling Globers – Episode 13 – Magnetic Declination" https://www.youtube.com/live/2BLIkqri2hg, 1:53:00-1:56:00.
49 Collamore, *His Pronouncement*, 45–46.
50 Ibid., 46.
51 Ibid.
52 Witsit Gets It, "Schooling Globers – Episode 13 – Magnetic Declination," https://www.youtube.com/live/2BLIkqri2hg, 1:04:20.
53 Ibid., 2:01:00–2:03:00.
54 Taboo Conspiracy, "How Time Zones Hide the Flat Earth by Hervé Riboni," https://www.youtube.com/watch?v=lMhheDWThxE, 2:00-4:00.
55 Flat Earth Sun Moon & Zodiac Clock App, "Whitsitt Gets Time Zones on a FLAT EARTH," https://www.youtube.com/watch?v=efzbYedUokg.
56 Taboo Conspiracy, "How Time Zones Hide the Flat Earth by Hervé Riboni," https://www.youtube.com/watch?v=lMhheDWThxE, 10:00.
57 Flat Earth Sun Moon & Zodiac Clock App, "Witsit Gets Time Zones on a FLAT EARTH," https://www.youtube.com/watch?v=efzbYedUokg.
58 Helmenstine, "Selenelion Eclipse," https://sciencenotes.org/selenelion-eclipse/.
59 Many of these videos in the playlist reference ideas pertinent to this whole section about eclipses.
60 MythologyWorldwide, "The Myth of the Ten Suns in Chinese Mythology," https://mythologyworldwide.com/the-myth-of-the-ten-suns-in-chinese-mythology/.
61 Nosachev, "The Influences of Western Esotericism on Russian Rock Poetry of the Turn of the Century," https://publications.hse.ru/pubs/share/direct/220857556.pdf.
62 These are the sorts of explanations and analogies often considered by Globe Skeptics.
63 Helfrich-Förster et al., "Women temporarily synchronize their menstrual cycles with the luminance and gravimetric cycles of the Moon," https://pubmed.ncbi.nlm.nih.gov/33571128/.
64 Hendrie, *The Greatest Lie on Earth*, 195–97.
65 Rowbotham, *Zetetic Astronomy*, 106–07.
66 Michael Tellinger, "A Time Before The Moon Was In The Sky View Larger Image," https://www.michaeltellinger.com/a-time-before-the-moon-was-in-the-sky/.

CHAPTER 6: SPACE AGENCIES

1 Flat Earth Sun Moon & Zodiac Clock App, "We haven't been to space U S S R Cosmonaut Igor Volk Confirmed," https://www.youtube.com/watch?v=Rs7MfWKSza0.

2 Ibid.
3 Young, *A Fool's Wisdom*, 54.
4 Kroupa and Haslbauer, "Our model of the universe has been falsified," https://iai.tv/articles/our-model-of-the-universe-has-been-falsified-auid-2393. The article states: "A recent publication in the journal Physical Review D with about 156 co-authors suggests the distribution of matter to be smoother than expected, based on the predictions of the standard model of cosmology."
5 Wall, "NASA gets $25.4 billion in White House's 2025 budget request," https://www.space.com/nasa-white-house-2025-budget-request.
6 Kimball, "The Outer Space Treaty at a Glance," https://www.armscontrol.org/factsheets/outer-space-treaty-glance.
7 See Stinnett, *Day of Deceit*, xiv. The book's afterword includes additional documentation that was not previously available in the original hardcover edition.
8 As cited in Paterson, "The Truth About Tonkin," https://www.usni.org/magazines/naval-history-magazine/2008/february/truth-about-tonkin. Also see Prados, "Essay: 40th Anniversary of the Gulf of Tonkin Incident," https://nsarchive2.gwu.edu/NSAEBB/NSAEBB132/essay.htm.
9 Memorandum for the Secretary of Defense, Unclassified, https://archive.org/details/OperationNorthwoods.
10 *Dispatch: Countering Criticism of the Warren Report*, January 4, 1967, https://history-matters.com/archive/jfk/cia/russholmes/104-10406/104-10406-10110/html/104-10406-10110_0002a.htm; National Archives, JFK Assassination Records, *Summary of Findings*, https://www.archives.gov/research/jfk/select-committee-report/summary.html.
11 Also see Gober, *An End to Upside Down Liberty*, 57–65, for examples. And Charles River Editors, *Project MK-Ultra: The History of the CIA's Controversial Human Experimentation Program.*
12 *A Structural Reevaluation of the Collapse of World Trade Center* 7, ii, https://files.wtc7report.org/file/public-download/A-Structural-Reevaluation-of-the-Collapse-of-World-Trade-Center-7-March 2020.pdf.
13 NPR, "Iraq WMD Timeline: How the Mystery Unraveled," https://www.npr.org/2005/11/15/4996218/iraq-wmd-timeline-how-the-mystery-unraveled.
14 NASA, "The NACA," https://www.nasa.gov/history/the-national-advisory-committee-for-aeronautics-naca/.
15 Science History Institute, *The Sex-Cult "Antichrist" Who Rocketed Us to Space*, https://www.sciencehistory.org/stories/disappearing-pod/the-sex-cult-antichrist-who-rocketed-us-to-space-part-1/.
16 Urban, *The Church of Scientology*, 40.
17 Fletcher, "The Nazis Had a Secret Project: Use Witchcraft to Make the Reich Last 1,000 Years," https://www.historynet.com/nazi-witches/.
18 See the documentary, *Out of Shadows*, https://www.outofshadows.org/.

19 PBS *Nova*, "A Tainted Legacy," https://www.pbs.org/wgbh/nova/sputnik/vonb-02.html.
20 As cited by Odle, *Like Clay Under the Seal*, 51–52.
21 Niilner, "Hitler Sent a Secret Expedition to Antarctica in a Hunt for Margarine Fat," https://www.history.com/news/hitler-nazi-secret-expedition-antarctica-whale-oil.
22 Austin Whitsitt has laid out helpful timelines that illustrate the Globe Skeptic perspective, such as his June 2024 Braveritas lecture, https://x.com/Witsitgetsit/status/1798165881429668322?s=46&t=ijTvU_628uOjAUiS9U8Hxw.
23 PBS *Nova*, "A Tainted Legacy," https://www.pbs.org/wgbh/nova/sputnik/vonb-07.html.
24 Ibid.
25 *American Moon*, 6:40–7:32, https://www.luogocomune.net/shop/films-in-english/american-moon-english-version.
26 Baron, "AN APOLLO REPORT," https://www.nasa.gov/history/Apollo204/barron.html.
27 *American Moon*, 17:30, https://www.luogocomune.net/shop/films-in-english/american-moon-english-version.
28 Howell, "Gus Grissom: 2nd American in Space," https://www.space.com/24443-gus-grissom.html.
29 For example, see Sibrel, *Moon Man*, 187.
30 Ibid., 188.
31 Griffith, "1 in 10 Americans Don't Believe the Moon Landing Really Happened," https://www.pcmag.com/news/1-in-10-americans-dont-believe-the-moon-landing-really-happened.
32 Sciencebusiness.net, "Conspiracy thinking: 25% of Europeans say the moon landing never happened," https://sciencebusiness.net/news/covid-19/conspiracy-thinking-25-europeans-say-moon-landing-never-happened.
33 *American Moon*, 25:00, https://www.luogocomune.net/shop/films-in-english/american-moon-english-version.
34 As stated in a *Britannica* article: "The Van Allen belts are most intense over the Equator and are effectively absent above the poles. No real gap exists between the two zones; they actually merge gradually, with the flux of charged particles showing two regions of maximum density. The inner region is centred approximately 3,000 km (1,860 miles) above the terrestrial surface. The outer region of maximum density is centred at an altitude of about 15,000 to 20,000 km (9,300 to 12,400 miles), though some estimates place it as far above the surface as six Earth radii (about 38,000 km [23,700 miles])," https://www.britannica.com/science/Van-Allen-radiation-belt.
35 *American Moon*, 1:03:18, https://www.luogocomune.net/shop/films-in-english/american-moon-english-version.
36 As cited in Ibid., 1:03:27.
37 Ibid., 1:06:00–1:07:00.
38 As cited in Ibid., 1:07:00.

39 NASA, "Apollo 8 Mission Report," https://www.nasa.gov/wp-content/uploads/static/history/alsj/a410/A08_MissionReport.pdf.
40 NASA, "Apollo 14 Mission Report," https://www.nasa.gov/wp-content/uploads/static/history/alsj/a14/A14MRntrs.pdf.
41 As shown in *American Moon*, 1:12:00–1:13:00, https://www.luogocomune.net/shop/films-in-english/american-moon-english-version.
42 As shown in Ibid., 1:09:50–1:11:00.
43 As shown in Ibid., 1:11:50.
44 As shown in Ibid.
45 thisisflatearth, "Don Pettit – I'd go to the Moon in a nanosecond [13]," https://rumble.com/v44jwbc-don-pettit-id-go-to-the-moon-in-a-nanosecond-.html.
46 Sibrel, *Moon Man*, 25–26.
47 Ibid., 30.
48 Ibid., 30–31.
49 Ibid., 32–33.
50 Ibid., 36.
51 Ibid., ch. 4.
52 Ibid., 62–63.
53 Ibid., 122.
54 Ibid., 76.
55 Bart Sibrel, "Moon Landing Fraud – Deathbed Confession – MM16," https://www.youtube.com/watch?v=wu5Z75ji3aU.
56 Sibrel, *Moon Man*, 117.
57 The Sol Foundation, "Iya Whiteley, Ph.D. on Trusting and Learning from Pilots," https://www.youtube.com/watch?app=desktop&v=HR09GHQ5AwA, 7:45–8:30.
58 See Russell Targ's statements at the beginning of *Where Is My Mind?* Ep. 4: CIA Psychic Spying & Knowing the Future.
59 The Sol Foundation, "Iya Whiteley, Ph.D. on Trusting and Learning from Pilots," https://www.youtube.com/watch?app=desktop&v=HR09GHQ5AwA, 8:30–9:10.
60 Sibrel, *Moon Man*, 23.
61 NASA, "The Apollo 11 Telemetry Data Recordings: A Final Report," 2, https://www.nasa.gov/wp-content/uploads/static/history/alsj/a11/apollo_11_tv_tapes_report.pdf.
62 As shown in *American Moon*, 1:36:00–1:40:00, https://www.luogocomune.net/shop/films-in-english/american-moon-english-version.
63 As shown in Ibid., 3:26:00–3:30:00.
64 Sibrel, *Moon Man*, 100.
65 Ibid., 105–06.
66 *The Telegraph*, https://www.telegraph.co.uk/news/science/space/6105902/Moon-rock-given-to-Holland-by-Neil-Armstrong-and-Buzz-Aldrin-is-fake.html.
67 Sibrel, *Moon Man*, 147.

68 Sullivan, "75-Year-Old Woman, Detained by NASA for Selling a Moon Rock, Can Sue," https://www.findlaw.com/legalblogs/ninth-circuit/75-year-old-woman-detained-by-nasa-for-selling-a-moon-rock-can-sue/.
69 As cited in Sibrel, *Moon Man*, 149.
70 Sibrel, *Moon Man*, 149.
71 Ibid., 137.
72 Sibrel, "NASA Admits Fraud – MM10–READ BELOW – Links 10 B-C-D," https://www.youtube.com/watch?v=x6WAuXjQDoo, 6:30-7:10.
73 Kranz, *Failure Is Not An Option*, 234–35.
74 Allen and Weaver, *The Apollo Moon Hoax*, 25.
75 As cited in Ibid.
76 Witsit Gets It, "NASA Fraud," https://www.youtube.com/live/CJnwl5aiBSk.
77 Clinton, *My Life*, 156.
78 Griffin, "Nasa releases first picture of the whole Earth taken in 43 years," https://www.independent.co.uk/tech/nasa-releases-first-picture-of-the-whole-earth-taken-in-43-years-10403944.html.
79 Jarrell, "Robert Simmon – AKA Mr. Blue Marble," https://www.nasa.gov/people-of-nasa/goddard-people/robert-simmon-aka-mr-blue-marble/.
80 Odle, *Like Clay Under the Seal*, 331–34. Also, see Darion, "Robert Simmon "Mr. Blue Marble," https://www.youtube.com/watch?v=u0I-H5VMBmY; and Flat Dave, "Finding the Curve–Einar Kuusk–FULL VERSION Parts 1–5," https://www.youtube.com/watch?v=B4N38jIW44w, 14:00–15:30.
81 Gietler, "Fisheye Lens vs. Wide-Angle Lens," https://www.uwphotographyguide.com/fisheye-lenses-underwater.
82 Smithsonian National Air and Space Museum, "Lens, 16mm, Fish-eye, Nikon, Shuttle," https://airandspace.si.edu/collection-objects/lens-16mm-fisheye-nikon-shuttle/nasm_A20080106000.
83 NASA Earth Observatory, "Fisheye Over Sinai," https://earthobservatory.nasa.gov/images/147993/fisheye-over-sinai.
84 Samuelson, "Take a Tour of the International Space Station in Dazzling 4K Detail," https://time.com/4549095/nasa-international-space-station-4k/.
85 NASA, "Training for Apollo 9," https://www.nasa.gov/image-article/training-apollo-9/.
86 Odle, *Like Clay Under the Seal*, 330.
87 raviprasad.ai Instagram account, https://www.instagram.com/reel/C8msOW4SbAB/?igsh=MXV4NGNoMTJ5OXhwbw==, June 24, 2024.
88 NASA Photojournal, https://photojournal.jpl.nasa.gov/mission/Galileo. The image itself is "PIA18176:Galileo Over Io (Artist's Concept)."
89 As cited in Odle, *Like Clay Under the Seal*, 397.

90 "Scientific Balloons," https://www.nasa.gov/scientificballoons/.
91 As cited in Odle, *Like Clay Under the Seal*, 388–89.
92 As cited in Ibid., 390–91.
93 As cited in Ibid., 387.
94 As cited in Ibid., 392.
95 As cited in Ibid., 414–15.
96 Andrews, "Common Issues with Satellite Phones: Troubleshooting Tips for Outdoor Adventurers," https://prepsurvivalright.com/common-issues-with-satellite-phones/.
97 Young, *A Fool's Wisdom*, 70.
98 Censored Important Videos, "April 2023: Justin Harvey calls out NASA at Brevard County," https://rumble.com/v2kkkmu-calling-out-nasa-april-2023.html.
99 Paul Davis UnCancelled, "Are the astronauts from the Challenger disaster still alive? Justin Harvey has the evidence," https://rumble.com/v52mvyw-are-the-astronauts-from-the-challenger-disaster-still-alive-justin-harvey-h.html.
100 Pearlman, "How did CBS' Big Bang Theory send Howard Wolowitz to space?" https://www.space.com/17891-big-bang-theory-space-set-design.html.
101 Witsit Gets It, "NASA Fraud," https://www.youtube.com/live/CJnwl5aiBSk, 2:59:30.
102 Ibid., 2:32:00.
103 Flat Earth Dave Archive, "Houston, We Have Got a HUGE Problem! ISS HOAX," https://www.youtube.com/watch?v=la6ec1ay0II.
104 Witsit Gets It, "NASA Fraud," https://www.youtube.com/live/CJnwl5aiBSk, 2:47:00-2:48:00.
105 A few examples shown in Witsit Gets It, "NASA Fraud," https://www.youtube.com/live/CJnwl5aiBSk, 2:47:00-2:53:00.
106 Moore et al., "Practical Applications of Cables and Ropes in the ISS Countermeasures System" https://ntrs.nasa.gov/api/citations/20160012382/downloads/20160012382.pdf.
107 Witsit Gets It, "NASA Fraud," https://www.youtube.com/live/CJnwl5aiBSk, 2:29:00–2:30:00.
108 Ibid., 3:15:20.
109 Witsit Gets It, "NASA Fraud – Episode 3," https://www.youtube.com/live/_gpM5lmRVW8, 2:13:00–2:16:00.
110 Additional bloopers shown at Witsit Gets It, "NASA Fraud," https://www.youtube.com/live/CJnwl5aiBSk, 3:19:00.
111 Witsit Gets It, "NASA Fraud–Episode 3," https://www.youtube.com/live/_gpM5lmRVW8.
112 Witsit Gets It, "NASA Fraud," https://www.youtube.com/live/CJnwl5aiBSk, 3:21:50.
113 NASA, "Nine Ways We Use AR and VR on the International Space Station," https://www.nasa.gov/missions/station/nine-ways-we-use-ar-and-vr-on-the-international-space-station/.

114 Witsit Gets It, "NASA Fraud," https://www.youtube.com/live/CJnwl5aiBSk, 3:36:50.
115 See the YouTube channel "Flat Earth and Coffee," https://www.youtube.com/@FlatEarthAndCoffee/videos.
116 NASA, "Neutral Buoyancy Laboratory," https://www.nasa.gov/johnson/neutral-buoyancy-laboratory/.
117 Pettit, "Panicked NASA cancels spacewalks after ISS astronaut's helmet 'fills with water,'" https://nypost.com/2022/05/19/panicked-nasa-cancels-spacewalks-after-iss-astronauts-helmet-fills-with-water/.
118 *Erin Burnett Out Front*, "NASA review underway after water leaks into astronaut's helmet," https://www.cnn.com/videos/business/2022/08/06/nasa-halts-spacewalks-spacesuit-problems-water-leak-astronaut-helmet-iss-fisher-pkg-ebof-vpx.cnn.
119 Also available on Rumble at the channel Don't Obey, "Bloopers From Space – Top 25 All Time Favorites," https://rumble.com/v28x1v0-bloopers-from-space-top-25-all-time-favorites.html.
120 Liverman v. McMurray, US Supreme Court, https://www.supremecourt.gov/DocketPDF/20/20-5808/176392/20210422160651080_20210422-160534-95753440-00001842.pdf, 2.

CHAPTER 7: THE IMPORTANCE OF CONSCIOUSNESS

1 Consciousness is difficult, if not impossible, to precisely define with words because it is not physical.
2 This topic is covered extensively in Gober, *An End to Upside Down Thinking*; and Gober, *An End to Upside Down Living* (among other books by Mark Gober). Many of the arguments also reflect summaries of much more detailed work by philosopher Bernardo Kastrup, PhD. For instance, see his book *The Idea of the World.*
3 Gosling, *Science and the Indian Tradition: When Einstein Met Tagore*, 162.
4 In addition to Kastrup, *The Idea of the* World, also see Kastrup's books *Brief Peaks Beyond* and *Why Materialism Is Baloney*. And see Gober, *An End to Upside Down Living*, 32–36.
5 Faggin, *Irreducible: Consciousness, Life, Computers, and Human Nature.*
6 See Kastrup, *Why Materialism Is Baloney*. Also see *Where Is My Mind?* podcast.
7 Young, *A Fool's Wisdom*, 247.
8 Ibid.
9 This analogy has been used by Kastrup and is also commonly used in arguments about germ vs. terrain theory. For example, see Gober, *An End to Upside Down Medicine.*
10 See Gober, *An End to Upside Down Thinking* and *Where Is My Mind?* podcast.
11 See *Where Is My Mind?* podcast, Ep. 2: Blindfolded.
12 Reville, "Remarkable story of maths genius who had almost no brain," https://www.irishtimes.com/news/

remarkable-story-of-maths-genius-who-had-almost-no-brain-1.1026845.

13 Pearsall, *The Heart's Code*, 7.
14 Compiled in Radin, *Real Magic*, 97.
15 Cardeña, "The experimental evidence for parapsychological phenomena: A review," https://doi.org/10.1037/amp0000236.
16 Utts, "An Assessment of the Evidence for Psychic Functioning," https://www.ics.uci.edu/~jutts/air.pdf. Also available in the *Journal of Parapsychology*.
17 See *Where Is My Mind?* podcast, Ep. 5: Near-Death Experiences, 13:20.
18 See Gober, *An End to Upside Down Thinking*, chs. 10 and 11. Also see the peer-reviewed research from the Windbridge Research Center, https://www.windbridge.org/research/.
19 University of Virginia's Division of Perceptual Studies, "Children Who Report Memories of Past Lives," https://med.virginia.edu/perceptual-studies/our-research/children-who-report-memories-of-previous-lives/.
20 Sharma and Tucker, "Cases of the Reincarnation Type with Memories from the Intermission Between Lives," https://med.virginia.edu/perceptual-studies/wp-content/uploads/sites/360/2015/11/Intermission-memories-JNDS.pdf.
21 Interview with Max Planck, *The Observer* (1931).
22 Schrödinger, *What Is Life? with Mind and Matter*, 139.
23 Radin, *Real Magic*, 197, 205.
24 This notion is often discussed by Rupert Spira, a spiritual philosopher.

CHAPTER 8: SPIRITUAL WAR

1 Barušs and Mossbridge, *Transcendent Mind*, 25.
2 Skeptiko, "Dr. Mario Beauregard, Frontier Science Wake Up Call |538|," https://www.youtube.com/watch?v=8i-Q0MTJFls&t=2114s.
3 Young, *A Fool's Wisdom*, ch. 5.
4 Ibid., 37–38. Also see Dr. Tom Cowan, "The Comparison Between Nuclear Physics and Virology – 11/8/23," https://www.bitchute.com/video/ZzCb07CEFbkp/. He discusses Dewey Larson's 1963 book *The Case Against the Nuclear Atom*.
5 A contention often made by cosmology researcher Austin Whitsitt.
6 As cited in Russell, *From Science to God*, 49.
7 Capra, *The Tao of Physics*, 203.
8 For instance, see Behe, *Darwin Devolves*; Meyer, *Return of the God Hypothesis*; and Dembski and McDowell, *Understanding Intelligent Design*.
9 Kastrup, *Brief Peeks Beyond*, 113–19. Also discussed in Gober, *An End to Upside Down Living*, 56–58.
10 See Christine Massey's website at https://www.fluoridefreepeel.ca/.
11 The Perth Group, "HIV—a virus like no other," https://www.theperthgroup.com/HIV/TPGVirusLikeNoOther.pdf.
12 See https://learninggnm.com/ for a primer on German New Medicine.

13 As cited in Radin, *Entangled Minds*, 73.
14 See Gober, *An End to Upside Down Contact*, 134–37.
15 Wahbeh, *The Science of Channeling*. For a further discussion of Wahbeh's work, see Gober, *An End to Upside Down Contact*.
16 See Gober, *An End to Upside Down Contact*. Also see Dennett, *The Healing Power of UFOs*.
17 Anita Moorjani, "The Day My Life Changed: February 2, 2006," https://www.anitamoorjani.com/my-nde.
18 Gallagher, *Demonic Foes*, 11.
19 Vincent, "Sex, rocket science, and Satanism: Meet Nasa's real hidden figures," https://www.telegraph.co.uk/films/2017/02/17/sex-rocket-science-scientology-meet-hidden-figures-behind-nasas/.
20 See, Gober, *An End to Upside Down Contact*, 57–66.
21 *Illinois Statutes Chapter 720*. Criminal Offenses §-33. Ritualized abuse of a child, https://codes.findlaw.com/il/chapter-720-criminal-offenses/il-stsect-720-5-12-33.html; *Idaho Statutes*, https://legislature.idaho.gov/statutesrules/idstat/Title18/T18CH15/SECT18-1506A/.
22 Davis, *Hell Minus One*, 85–93 (Kindle).
23 Ibid., 1858–1863 (Kindle).
24 "RAW: Sean Stone with Jay Parker," https://robertdavidsteele.com/sean-stone-with-jay-parker/.
25 *SATAN'S CHILDREN – SATANIC RITUAL ABUSE (60 MINUTES AUSTRALIA)*, https://www.bitchute.com/video/I1bHD2JyfmEg/.
26 REALWOMEN/REALSTORIES, (UNCUT) Anneke Lucas: Sold by my mother into a pedophile network at six years of age, https://www.youtube.com/watch?v=RTxD6VUcxZ0. Also see 58 Tsakiris, Anneke Lucas, Recovering From Unimaginable Evil |441|, https://skeptiko.com/anneke-lucas-recovering-from-unimaginable-evil-441/.
27 "RAW: Sean Stone with Jay Parker," https://robertdavidsteele.com/sean-stone-with-jay-parker/.
28 This is a common discussion point among Globe Skeptics.
29 See the work of Dr. Ardy Sixkiller Clarke. For example, *Sky People: Untold Stories of Alien Encounters in Mesoamerica* and *More Encounters with Star People: Urban American Indians Tells Their Stories* and *Space Age Indians: Their Encounters with the Blue Men, Reptilians, and Other Star People*.
30 US Department of Homeland Security, "Kona Blue" (presentation), PDF pp. 18, 29, https://www.aaro.mil/Portals/136/PDFs/UAP_RECORDS_RESEARCH/AARO_DHS_Kona_Blue.pdf?ver=BjOpTzFISPc0LWMw5uAzzw%3d%3d.
31 Swedenborg, *Other Planets*, 3.
32 Michael Salla, https://michaelsalla.com/.
33 Richard Kositzke, "Secret Space Program & Flat Earth Dome /w Aug Tellez [Strongholds]," https://www.youtube.com/watch?v=kH6wFf0eMWU, 1:29:00-1:32:00.
34 Vallee, *Revelations*, 237. Vallee also suggests that the beings could be

from "galaxies," but the point here is that, contrary to much of modern UFO thinking, that doesn't necessarily have to be the case.

35 Consider a number of examples: In Hindu mythology, Patala is one of the seven realms beneath the Earth, inhabited by serpents that can shapeshift into humans called *Nagas*; in Buddhist traditions, Shambhala is considered a hidden paradise often associated with underground realms; in Greek mythology, the underworld where souls of the dead reside is ruled by Hades; in Norse mythology, there is Hel, the realm of the dead, and Svartalfheim, an underground world of dwarves; in Hopi (Native American) lore, Sipapu is a portal to the underworld; and in ancient Mesopotamian lore, Kur was a dark underground world of the dead.

36 Cooper, Blumenthal, and Kean, "Glowing Auras and 'Black Money': The Pentagon's Mysterious U.F.O. Program," https://www.nytimes.com/2017/12/16/us/politics/pentagon-program-ufo-harry-reid.html. Also see Blumenthal, "On the Trail of a Secret Pentagon U.F.O. Program," https://www.nytimes.com/2017/12/18/insider/secret-pentagon-ufo-program.html.

37 Vallee, *Revelations*, 236–37.

38 Dr. Steven Greer, "Von Braun's Legacy (with Dr. Carol Rosin)," https://www.youtube.com/watch?v=gP8ftWzFYI4.

39 Vallee and Davis, "Incommensurability, Orthodoxy and the Physics of High Strangeness: A 6-layer Model for Anomalous Phenomena," 8, https://www.jacquesvallee.net/wp-content/uploads/2018/11/Incommensurability_Orthodoxy_and_the_Phy.pdf.

40 Bishop, *Project Beta*, 8.

41 See Gober, *An End to Upside Down Contact*, 196.

42 See Ibid., 127.

43 See Ibid., 46.

44 These topics are covered in greater detail in Gober, *An End to Upside Down Contact*.

45 Biglino, *Gods of the Bible*, back cover.

46 Ibid., 58.

47 Ibid.

48 Wallis, *Escaping from Eden*, 144.

49 For instance, see Robinson, preface in *The Nag Hammadi Scriptures*, xi–xii.

50 "The Secret Book of John" in *The Nag Hammadi Scriptures*, 108.

51 Ibid., 110.

52 Ibid., 116.

53 Ibid., 117.

54 Ibid., 119.

55 Ibid.

56 Ibid., 125.

57 Ibid., 194.

58 "On the Origin of the World," in the *Nag Hammadi Scriptures*, 221.

59 For example, see Gober, *An End to Upside Down Contact*, 72.
60 For example, see Lee, *The One World Tartarians*. Also see the 2024 documentary *Old World Order*, https://owofilm.com/.
61 The link to Mike Wilkerson's YouTube channel is https://www.youtube.com/@stellium7.

CONCLUSION: FREEDOM THROUGH TRUTH

1 See Jahn and Dunne, *Consciousness and the Source of Reality*. For more on random-number generator studies, see Bosch, Steinkamp & Boller, "Examining psychokinesis: The interaction of human intention with random number generators—a meta-analysis; Radin, Nelson, Dobyns, & Houtkooper, "Re-examining psychokinesis: Commentary on the Bösch, Steinkamp and Boller metaanalysis."
2 See https://global-mind.org/. Also see Nelson, Radin, Shoup, and Bancel, "Correlations of continuous random data with major world events."
3 New Thinking Allowed Foundation, "The Mandela Effect with Cynthia Sue Larson." New Thinking Allowed Foundation website, https://www.newthinkingallowed.org/the-mandela-effect-with-cynthia-sue-larson/, 7:30-7:40.
4 For example, see MoneyBags73, "Mandela Effect Passenger Airliners Look Different In This Reality!! Voting Video #248," https://www.youtube.com/watch?v=IBw1CkU5Xeo.
5 For instance, see Liles, "50 Mandela Effect Examples of Things You *Think* You Remember Correctly (That You've Actually Got All Wrong)," https://parade.com/1054775/marynliles/mandela-effect-examples/; Snow, "The Mandela Effect and the Bible," https://www.artesianministries.org/faith/the-mandela-effect-and-the-bible/; Hazel Fiver, "Aircraft Engines Under Wings – Mandela Effect," https://www.youtube.com/watch?v=GrvOhdRBrLI.
6 See Mossbridge, Tressoldi & Utts, "Predictive physiological anticipation preceding seemingly unpredictable stimuli: A meta-analysis"; Bem, Tressoldi, Raberyon, & Duggan, "Feeling the future: A meta-analysis of 90 experiments on the anomalous anticipation of random future events."
7 See Gober, *An End to Upside Down Thinking*, ch. 6.
8 Spivak and Saunders, *An Antidote to Violence*. The quotation is from the Amazon.com description: https://www.amazon.com/Antidote-Violence-Evaluating-Evidence/dp/1789042585.

BIBLIOGRAPHY

Abizaid, John G. *The Enlightenment of the World.* Boston, 1910.

Aerostudents.com. "Flight Dynamics Summary." n.d. https://www.aerostudents.com/courses/flight-dynamics/flightDynamicsFullVersion.pdf.

Aether Cosmology. "Downward Acceleration – Gravity Anomalies – Thunderstorms." *Aether Cosmology* website, May 3, 2024. https://aethercosmology.com/t/downward-acceleration-gravity-anomalies-thunderstorms/134.

Aether Cosmology. "Long-Distance Specular Reflections Illustrate a Lack of Curvature." Aether Cosmology website, March 17, 2024. https://aethercosmology.com/t/long-distance-specular-reflections-illustrate-a-lack-of-curvature/95.

Albrecht, Andreas et al. "Report of the Dark Energy Task Force." September 20, 2006. https://arxiv.org/pdf/astro-ph/0609591.

Aldrin, Colonel Edwin E. "Buzz," and Wayne Warga. *Return to Earth.* New York: Random House, 1973.

Alencar, Eddie. *16 Emergency Landings Proving Flat Earth.* Eddie Alencar, 2019.

Allen, Marcus, and Trevor Weaver. *The Apollo Moon Hoax: The Real Evidence—A Reference Guide to the Facts.* 2023.

American Oceans. "How Do Whales Communicate?" American Oceans website, n.d. https://www.americanoceans.org/facts/how-do-whales-communicate/.

Andrews, Chris. "Common Issues with Satellite Phones: Troubleshooting Tips for Outdoor Adventurers." Prep Survival Right, May 29, 2024. https://prepsurvivalright.com/common-issues-with-satellite-phones/.

Anti-Disinfo Leauge [*sic*]. "True Earth 101: Curvature." BitChute video, 2023. https://www.bitchute.com/video/6nbYsiGpwAH9/.

Arctic Council. "Protected areas in the Arctic region." Arctic Council's website, n.d. https://arctic-council.org/news/protected-areas-in-the-arctic-region/.

Assis, André. *Relational Mechanics*. Montreal: Apeiron, 1999.

Assis, André et al. "Hubble's Cosmology: From a Finite Expanding Universe to a Static Endless Universe." *arXiv*, 2011. https://arxiv.org/pdf/0806.4481.

Ault, Alicia. "How Does Foucault's Pendulum Prove the Earth Rotates?" *Smithsonian Magazine* website, February 2, 2018. https://www.smithsonianmag.com/smithsonian-institution/how-does-foucaults-pendulum-prove-earth-rotates-180968024/.

Bailey, Mark. "A Farewell to Virology" (Expert Edition). Dr. Sam Bailey website, September 15, 2022. https://drsambailey.com/a- farewell-to-virology-expert-edition/.

Bailey, Mark, and John Bevan-Smith. "The COVID-19 Fraud & War on Humanity." Dr. Sam Bailey website, November 11, 2021. https://drsambailey.com/the-covid-19-fraud-war-on-humanity/

Baker, Adolf. *Modern Physics & Antiphysics*. Reading, MA: Addison-Wesley, 1967.

Barnett, Lincoln. *The Universe and Dr. Einstein*. Garden City: Dover Publications, 2005.

Baron, Thomas Ronald. "An Apollo Report." *NASA* website, n.d. https://www.nasa.gov/history/Apollo204/barron.html.

Barr, Roseanne. "What's Beyond Antarctica? with Flat Earth Dave | The Roseanne Barr Podcast #55." YouTube video, July 4, 2024. https://www.youtube.com/watch?v=bIeOPtHY4KI.

Baruŝs, Imants, and Julia Mossbridge. *Transcendent Mind: Rethinking the Science of Consciousness*. Washington, DC: American Psychological Association, 2017.

BBC. "Astronomer Tycho Brahe 'not poisoned,' says expert." BBC website, November 15, 2012. https://www.bbc.com/news/science-environment-20344201.

Beckley, Timothy Green, and Tim R. Swartz. *The Missing Diary of Admiral Richard E. Byrd*. New Brunswick: Global Communications, 2013.

Behe, Michael J. *Darwin Devolves*. New York: HarperOne, 2019.

Beischel, J., Boccuzzi, M., Biuso, M., and Rock, A. J. (2015). "Anomalous information reception by research mediums under blinded conditions II: Replication and extension." *EXPLORE: The Journal of Science & Healing*, 11(2), 136–142. doi: 10.1016/j.explore.2015.01.001.

Bem, Daryl et al. "Feeling the Future: A Meta-Analysis of 90 Experiments on the Anomalous Anticipation of Random Future Events." *F1000 Research* 4 (2015): 1188. https://f1000research.com/articles/4-1188/v1.

Bennett, Bo. *Logical Fallacies: The Ultimate Collection of Over 300 Logical Fallacies*. Sudbury: Archieboy Holdings, 2021.

Bertone, Gianfranco, and Dan Hooper. "History of dark matter." *Rev. Mod. Phys.*, July 6, 2018. https://link.aps.org/accepted/10.1103/RevModPhys.90.045002.

Biglino, Mauro. *Gods of the Bible*. Torino: Tuthi, 2023.

Bikos, Daniel, and Aparna Kher. "Twilight, Dawn, and Dusk." *Time and Date* website, n.d. https://www.timeanddate.com/astronomy/different-types-twilight.html.

Bishop, Greg. *Project Beta: The Story of Paul Bennewitz, National Security, and the Creation of a Modern UFO Myth*. New York: Paraview Pocket Books, 2005.

Blumenthal, Ralph. "On the Trail of a Secret Pentagon U.F.O. Program." *New York Times*, December 18, 2017. https://www.nytimes.com/2017/12/18/insider/secret-pentagon-ufo-program.html.

Bodans, David. *Electric Universe: The Shocking True Story of Electricity*. New York: Crown, 2005.

Boll, Mike. "Neil DeGrasse Tyson Destroys the Eratosthenes Story." YouTube video, May 5, 2018. https://www.youtube.com/watch?v=Gs0M0FJbKI4.

Bosch, Holger, Fiona Steinkamp, and Emil Boller. "Examining Psychokinesis: The Interaction of Human Intention with Random Number Generators; A Meta-Analysis." *Psychological Bulletin* 132, no. 4 (2006): 497–523.

Britannica. "Aristarchus of Samos." *Britannica* website, n.d. https://www.britannica.com/biography/Aristarchus-of-Samos.

Britannica. "dark matter." *Britannica* website, n.d. https://www.britannica.com/science/dark-matter.

Britannica. "Ether." *Britannica* website, n.d. https://www.britannica.com/science/ether-theoretical-substance.

Britannica. "International Geophysical Year." *Britannica* website, n.d. https://www.britannica.com/event/International-Geophysical-Year.

Britannica. "Michelson-Morley Experiment." *Britannica* website, n.d. https://www.britannica.com/science/Michelson-Morley-experiment.

Britannica. "Ptolemaic system." *Britannica* website, n.d. https://www.britannica.com/science/Ptolemaic-system

Britannica. "Relativity." *Britannica* website, n.d. https://www.britannica.com/science/relativity.

Britannica. "Tychonic system." *Britannica* website, n.d. https://www.britannica.com/biography/Tycho-Brahe-Danish-astronomer.

Britannica. "Van Allen radiation belt." Britannica website, n.d. https://www.britannica.com/science/Van-Allen-radiation-belt.

Byrd Jr., Richard Evelyn. *Discovery: The Story of the Second Byrd Antarctic Expedition.* New York: Rowman and Littlefield, 2015.

Byrd Jr., Richard Evelyn. *Exploring with Byrd: Episodes of an Adventurous Life.* New York: Rowman and Littlefield, 2015.

Cadbury, Deborah. *Space Race.* New York: Harper Perennial, 2006.

Capra, Fritjof. *The Tao of Physics: An Exploration of the Parallels between Modern Physics and Eastern Mysticism.* Boston: Shambhala, 1975.

Cardeña, Etzel. "The experimental evidence for parapsychological phenomena: A review." *American Psychologist*, 73(5) (2018), 663–77. https://doi.org/10.1037/amp0000236.

Carpenter, WM. *One Hundred Proofs That the Earth Is Not a Globe.* Nouveau Classics, 2017.

Caspar, Max. *Kepler*. New York: Dover Publications, 1993.

Castro, Joseph. "Which Direction Does Toilet Water Swirl at the Equator?" *Live Science* website, October 25, 2011. https://www.livescience.com/33567-toilet-swirl-direction-equator.html.

Cattaneo, Giorgio. *Mauro Biglino: The Naked Bible, The Truth about the Most Famous Book in History.* Torino: Tuthi, 2021.

CBC News. "NORAD Upgrades a 'Good Move' for Northern Security, Says Nunavut MP." CBC News website, June 22, 2022. https://www.cbc.ca/news/canada/north/norad-upgrade-defence-nunavut-nws-1.6498070.

Celebrate Truth. "Rob Skiba vs. Dr. Robert Sungenis | Flat Earth Biblical Debate 2018." YouTube video, November 28, 2018. https://www.youtube.com/watch?v=AqbiwtRKrtg.

Censored Important Videos. "April 2023: Justin Harvey calls out NASA at Brevard County." Rumble video, April 26, 2023. https://rumble.com/v2kkkmu-calling-out-nasa-april-2023.html.

Chaikin, Andrew. *A Man on the Moon*. Great Britain: Penguin Books, 2019.

Charan, Radha. *Book 1: Vedic Universe: Flat Earth.* 2018.

Charles River Editors, *Project MK-Ultra: The History of the CIA's Controversial Human Experimentation Program.* Independently Published, 2022.

Chief, Woview. "Dispatch: Countering Criticism of the Warren Report." January 4, 1967, https://history-matters.com/archive/jfk/cia/russholmes/104-10406/104-10406-10110/html/104 10406-10110_0002a.htm.

CIA. "CIA-RDP86-00513R00134372008-3." Approved for release June 15, 2000. https://www.cia.gov/readingroom/docs/CIA-RDP86-00513R001343720008-3.pdf.

Clark, Ardy Sixkiller. *More Encounters with Star People: Urban American Indians Tell Their Stories*. San Antonio, TX: Anomalist Books, 2016.

Clark, Ardy Sixkiller. *Space Age Indians: Their Encounters with Blue Men, Reptilians, and Other Star People*. San Antonio, TX: Anomalist Books, 2019.

Clark, Ronald W. *Einstein: The Life and Times*. New York: Harper Perennial, 2007.

Clarke, Arthur C. *Profiles of the Future: An Inquiry into the Limits of the Possible*. New York: Harper & Row, 1962.

Clinton, Bill. *My Life*. New York: Knopf Publishing Group, 2004.

Cohen, I. Bernard. *The Birth of a New Physics*. New York: W.W. Norton & Company, 1985. https://archive.org/details/B-001-015-439/page/n95/mode/2up?q=%22prove+that+the+earth%22.

Coleman, James A. *Relativity for the Layman*. New York: Signet Books, 1958.

Collamore, Robert Gould Shaw. *His Pronouncement*. Philadelphia: Dorrance & Company, 1924.

Confunden, Nos. *Hidden Lands Beyond the Antarctica*. Claudio Nocelli, 2023.

Confunden, Nos. *The Lands of Mars*. Claudio Nocelli, 2022.

Confunden, Nos. *The Navigator Who Crossed the Ice Walls*. Nos Confunden, 2022.

Confunden, Nos. *Terra-Infinita*. Claudio Nocelli, 2022.

Coomes, Tom. "Mirage of Chicago Skyline Seen from Michigan Shoreline." ABC57 News website, April 29, 2015. https://www.abc57.com/news/mirage-of-chicago-skyline-seen-from-michigan-shoreline.

Cooper, Gene. "Adding Liquid Payloads Effects to the 6-DOF Trajectory of Spinning Projectiles." Army Research Laboratory, March 2010. https://apps.dtic.mil/sti/pdfs/ADA519118.pdf.

Cooper, Helene, Ralph Blumenthal, and Leslie Kean. "Glowing Auras and 'Black Money': The Pentagon's Mysterious U.F.O. Program," *New York Times*, December 16, 2017. https://www.nytimes.com/2017/12/16/us/politics/pentagon-program-ufo-harry-reid.html.

Copernicus, Nicolaus. *On the Revolution of Heavenly Spheres*. Amherst: Prometheus Books, 1995.

Cowan, Dr. Tom. "The Comparison Between Nuclear Physics and Virology – 11/8/23." Bitchute video, November 8, 2023. https://www.bitchute.com/video/ZzCb07CEFbkp/.

Cowan, Dr. Tom. "Q&A/ 'Do Viruses Exist?' Article Webinar from May 29th, 2024." Rumble video, May 29, 2024. https://rumble.com/v4y9xlr-q-and-a-do-viruses-exist-article-webinar-from-may-29th-2024.html?

Cowan, Thomas. *Breaking the Spell: The Scientific Evidence for Ending the COVID Delusion*. Thomas S. Cowan, MD, 2021.

Cowan, Thomas, and Sally Fallon Morell. *The Truth about Contagion: Exploring Theories of How Disease Spreads* (previously titled *The Contagion Myth*). New York: Skyhorse Publishing, 2021.

Cravat, Marcelina, and Andrew Kaufman, MD. *TERRAIN: The Film*. 2022, https://terrainthefilm.com/.

Crrow777Radio. *Shoot the Moon*. Documentary, April 5, 2019. Vimeo. https://vimeo.com/ondemand/shootthemoon.

Darion. "Robert Simmon 'Mr. Blue Marble.'" YouTube video, June 26, 2019. https://www.youtube.com/watch?v=u0I-H5VMBmY.

Davidson, Thomas. "Free Energy, Gravity and the Aether." October 18, 1997. http://tuks.nl/pdf/Reference_Material/Paul_Stowe/Free%20Energy,%20Gravity%20and%20the%20Aether%20101897.pdf.

DeCamp, John W. *The Franklin Cover-Up: Child Abuse, Satanism, and Murder in Nebraska*. Lincoln, NE: AWT, Inc., 2006 (original version, 1992).

Dembski, William A., and Sean McDowell. *Understanding Intelligent Design*. Eugene: Harvest House Publishers, 2008.

Dennett, Preston. *The Healing Power of UFOs: 300 True Accounts of People Healed by Extraterrestrials*. Self-published, 2019.

DITRH. *Meteors Determine Up and Down*. YouTube video, May 12, 2020. https://www.youtube.com/watch?v=RhQHV0AAFrU.

Division of Perceptual Studies. "Children Who Remember Previous Lives – Academic Publications." University of Virginia, Division of Perceptual Studies website, n.d. https://med.virginia.edu/perceptual-studies/publications/academic-publications/children-who-remember-previous-lives-academic-publications/.

Don't Obey. "Bloopers from Space – Top 25 All-Time Favorites." Rumble video, February 9, 2023. https://rumble.com/v28x1v0-bloopers-from-space-top-25-all-time-favorites.html.

Dubay, Eric. *Flatlantis*. Eric Dubay, 2020.

Dubay, Eric. *The Flat-Earth Conspiracy*. Eric Dubay, 2014.

Dubay, Eric. *200 Proofs Earth Is Not a Spinning Ball*. Eric Dubay, 2015.

Duke, E. L. et al. "Derivation and Definition of a Linear Aircraft Model." *NASA*, 1988. https://ntrs.nasa.gov/api/citations/19890005752/downloads/19890005752.pdf.

Dunham, Ella. "4 Reasons Why You Can't Fly Over Antarctica (and 4 Exceptions)." *Executive Flyers* website, September 7, 2023. https://executiveflyers.com/why-cant-you-fly-over-antarctica/.

Edge. "The Energy of Empty Space That Isn't Zero: A Talk with Lawrence Krauss." Edge website, n.d. https://www.edge.org/conversation/the-energy-of-empty-space-that-isn-39t-zero.

Edwards, Philip (Ed.). *James Cook: The Journals*. London: Penguin Books, 1999.

Einstein, Albert. "How I Created the Theory of Relativity." *Physics Today*, August 1982. https://courses.physics.ucsd.edu/2022/Fall/physics2d/einsteinonrelativity.pdf.

Einstein, Albert. Relativity: *The Special and General Theory*. New York: Henry Holt and Company, 1920. https://archive.org/details/relativitythespe30155gut.

Einstein, Albert. "To Ernst Mach." *Volume 5: The Swiss Years: Correspondence, 1902–1914 (English translation supplement)*. June 25, 1913, Doc. 448, p. 340. https://einsteinpapers.press.princeton.edu/vol5-trans/362

Einstein, Albert, and Leopold Infeld. *The Evolution of Physics: From Early Concepts to Relativity and Quanta*. New York: Simon and Schuster, 1938. https://archive.org/details/evolutionofphysi033254mbp.

Emerson, Willis George. *The Smoky God: A Voyage to the Inner World*. Las Vegas: Lits, 2009.

EndeavorFreedom. "Rob Skiba and Zen Garcia – FE Court Case and the Great Deception." YouTube video, July 29, 2019. https://www.youtube.com/watch?v=lv2-T3nKjb0.

The End of COVID. n.d. https://theendofcovid.com/.

Erin Burnett Out Front. "NASA review underway after water leaks into astronaut's helmet." CNN video, August 6, 2022. https://www.cnn.com/videos/business/2022/08/06/nasa-halts-spacewalks-spacesuit-problems-water-leak-astronaut-helmet-iss-fisher-pkg-ebof-vpx.cnn.

Everyday Sounds. "Ken Campbell – Reality on the Rocks – Part 3 – Beyond our Ken." YouTube video, May 26, 2023. https://www.youtube.com/watch?v=6Qf0z5ZMjJs.

Faggin, Federico. *Irreducible: Consciousness, Life, Computers, and Human Nature.* Alresford: Essentia Books, 2024.

Faith-Masterson, Serena. *I Am Serena.* Dream Magic Publications, 2020.

Feynman, Richard P. "Electricity in the Atmosphere." *The Feynman Lectures on Physics,* Vol. II, 1963. https://www.feynmanlectures.caltech.edu/II_09.html.

FindLaw for Legal Professionals, "Illinois Statutes Chapter 720. Criminal Offenses §-33. Ritualized abuse of a child." FindLaw website, n.d. https://codes.findlaw.com/il/chapter-720-criminal-offenses/il-st-sect-720-5-12-33.html.

Flat Dave. "Finding the Curve – Einar Kuusk – FULL VERSION Parts 1–5." YouTube video, May 12, 2024. https://www.youtube.com/watch?v=B4N38jIW44w.

Flat Earth and Coffee. "Flat Earth and Coffee Videos." YouTube channel, n.d. https://www.youtube.com/@FlatEarthAndCoffee/videos.

Flat Earth, Banjo, USA, Japan, and Brazil. "Falling Stars Prove FLAT EARTH." YouTube video, April 6, 2022. https://www.youtube.com/watch?v=Ei8lfNHus1M.

Flat Earth Dave Archive. "Houston, We Have Got a HUGE Problem! ISS HOAX." YouTube video, February 20, 2024. https://www.youtube.com/watch?v=la6ec1ay0II.

Flat Earth Dave Interviews. "Brad Olson Discusses Flat Earth but Remains Lost in Antarctica." YouTube video, May 14, 2023. https://youtu.be/IsWJWvSKrac?si=jCs7AU7DU9QBMSgV.

Flat Earth Dave Interviews 2. "SEASONS EXPLAINED on a Flat Earth – Justin Peters Ministries." YouTube video, March 25, 2024. https://www.youtube.com/watch?si=GD_qdr5flZyanXwL&v=ggzJnG9ERjM&feature=youtu.be.

Flat Earth Sun, Moon & Zodiac Clock App. "Antarctica 24 Hour Sun FAKED Again!" YouTube video, February 24, 2024. https://www.youtube.com/watch?v=o7uLeVKv0Sc.

Flat Earth Sun, Earth Moon & Zodiac Clock App. "Humpback Whales Prove Our Flat Earth." YouTube video, December 12, 2019. https://www.youtube.com/watch?v=KIjO1_KCXJA.

Flat Earth Sun, Moon & Zodiac Clock App. "Meteorite Crater or Guizer?" YouTube video, March 5, 2022. https://www.youtube.com/watch?si=nY5vjwFe_fMcyeAY&v=Tq7MFz_aQog&feature=youtu.be.

Flat Earth Sun, Moon & Zodiac Clock App. "Meteors and Meteor Showers." YouTube video, March 5, 2022. https://www.youtube.com/watch?si=4EMy1qbZliSGxCoB&v=Ir-8rptN-k8&feature=youtu.be.

Flat Earth Sun, Moon & Zodiac Clock app. "NETFLIX Debunks Flat Earth – Ring Laser Gyro and Laser Experiment Deception." YouTube video, October 17, 2023. https://youtu.be/XHh2X_dWdxA?si=CuDGvdiGKjlJm-hL.

Flat Earth Sun, Moon & Zodiac Clock App. "We Haven't Been to Space U S S R Cosmonaut Igor Volk Confirmed." YouTube video, May 19, 2023. https://www.youtube.com/watch?v=Rs7MfWKSza0.

Flat Earth Sun, Moon & Zodiac Clock App. "Whitsitt Gets Time Zones on a FLAT EARTH." YouTube video, June 9, 2023. https://www.youtube.com/watch?v=efzbYedUokg.

FlatEarth.ws. "Negative Parallax." FlatEarth.ws website, n.d. https://flatearth.ws/negative-parallax.

Flatsmack Films. *Flatlanders*. n.d. https://flatsmackfilms.com/flatlanders.

Fletcher, Zita Ballinger. "The Nazis Had a Secret Project: Use Witchcraft to Make the Reich Last 1,000 Years." HistoryNet website, October 6, 2022. https://www.historynet.com/nazi-witches/.

Fluoride Free Peel website. https://www.fluoridefreepeel.ca/.

France24. "To 11 Million Brazilians, the Earth Is Flat." France 24 website, February 28, 2020. https://www.france24.com/en/20200228-to-11-million-brazilians-the-earth-is-flat.

Freedman, Ethan. "What Percentage of the Ocean Have We Mapped?" *Live Science* website, October 24, 2023. https://www.livescience.com/planet-earth/rivers-oceans/what-percentage-of-the-ocean-have-we-mapped.

Gallagher, Richard. *Demonic Foes*. New York: HarperOne, 2022.

Garcia, Zen. *The Flat Earth: As Key to Decrypt the Book of Enoch*. Zen Garcia, 2015.

Garwood, Christine. *Flat Earth: The History of an Infamous Idea.* London: Pan Macmillan, 2007.

Giannini, F. Amadeo. *Worlds Beyond the Poles*. Theonomos, 1959.

Gietler, Scott. "Fisheye Lens vs. Wide-Angle Lens." *Underwater Photography Guide* website, n.d. https://www.uwphotographyguide.com/fisheye-lenses-underwater.

Gilder, Joshua, and Anne-Lee Gilder. *Heavenly Intrigue: Johannes Kepler, Tycho Brahe, and the Murder Behind One of History's Greatest Scientific Discoveries*. New York: Doubleday, 2004

The Global Consciousness Project. "Formal Results: Testing the GCP Hypothesis." Global Consciousness Project website, n.d. global-mind.org/results.html.

Gober, Mark. *An End to Upside Down Contact: UFOs, Aliens, and Spirits—and Why Their Ongoing Interaction with Human Civilization Matters*. Cardiff-by-the-Sea, CA: Waterside Productions, 2022.

Gober, Mark. *An End to Upside Down Liberty: Turning Traditional Political Thinking on Its Head to Break Free from Enslavement.* Cardiff-by-the-Sea, CA: Waterside Productions, 2021.

Gober, Mark. *An End to Upside Down Living: Reorienting Our Consciousness to Live Better and Save the Human Species*. Cardiff by-the-Sea, CA: Waterside Productions, 2020.

Gober, Mark. *An End to Upside Down Medicine: Contagion, Viruses, and Vaccines—and Why Consciousness Is Needed for a New Paradigm of Health*. Cardiff-by-the-Sea, CA: Waterside Publishing, 2023.

Gober, Mark. *An End to Upside Down Thinking: Dispelling the Myth That the Brain Produces Consciousness, and the Implications for Everyday Life*. Cardiff-by-the-Sea, CA: Waterside Publishing, 2018.

Gober, Mark. *An End to the Upside Down Reset: The Leftist Vision for Society Under the Great Reset—and How It Can Fool Caring People into Supporting Harmful Causes*. Cardiff-by-the-Sea, CA: Waterside Productions, 2023.

Gober, Mark, and Blue Duck Media. *Where Is My Mind?* Podcast (Eps. 1–8). June–September 2019.

Gohd, Chelsea. "What Is Dark Energy? Inside Our Accelerating, Expanding Universe." *NASA Science* website, n.d. https://science.nasa.gov/universe/the-universe-is-expanding-faster-these-days-and-dark-energy-is-responsible-so-what-is-dark-energy/.

Gosling, David. *Science and the Indian Tradition: When Einstein Met Tagore*. London: Routledge, 2007.

Gould, Stephen Jay. *Galileo Galilei: Dialogue Concerning the Two Chief World Systems*. New York: The Modern Library, 2021.

Graham, David A. "Rumsfeld's Knowns and Unknowns: The Intellectual History of a Quip." *The Atlantic* website, March 27, 2014. https://www.theatlantic.com/politics/archive/2014/03/rumsfelds-knowns-and-unknowns-the-intellectual-history-of-a-quip/359719/.

Greer, Steven. "Von Braun's Legacy (with Dr. Carol Rosin)." YouTube video, October 11, 2013. https://www.youtube.com/watch?v=gP8ftWzFYI4.

Griffin, Andrew. "Nasa Releases First Picture of the Whole Earth Taken in 43 Years." *The Independent* website, July 21, 2015. https://www.independent.co.uk/tech/nasa-releases-first-picture-of-the-whole-earth-taken-in-43-years-10403944.html.

Griffith, Erin. "1 in 10 Americans Don't Believe the Moon Landing Really Happened." *PCMag* website, July 18, 2019. https://www.pcmag.com/news/1-in-10-americans-dont-believe-the-moon-landing-really-happened.

Hamer, John. *The Falsification of Science: Our Distorted Reality.* Self-published, 2021.

Hannam, James. "A Square, Flat Earth and Round Heaven Meant the World to Ancient China." *Atlas Obscura* website, September 21, 2023. https://www.atlasobscura.com/articles/flat-earth-ancient-china.

Hawking, Stephen. *A Brief History of Time*. New York: Bantam Books, 1988.

Hawking, Stephen, and Leonard Mlodinow. *The Grand Design*. New York: Bantam Books, 2010.

Hazel Fiver. "Aircraft Engines Under Wings – Mandela Effect." YouTube video, September 7, 2020. https://www.youtube.com/watch?v=GrvOhdRBrLI.

Helfrich-Förster, Charlotte et al. "Women Temporarily Synchronize Their Menstrual Cycles with the Luminance and Gravimetric Cycles of the Moon." *Science Advances* 7, no. 5 (2021). https://pubmed.ncbi.nlm.nih.gov/33571128/.

Helmenstine, Anne. "Selenelion Eclipse." *Science Notes*, June 1, 2022. https://sciencenotes.org/selenelion-eclipse/.

Hendrie, Edward. *The Greatest Lie on Earth: Proof That Our World Is Not a Moving Globe.* Great Mountain Publishing, 2016.

Hickson, Gerrard. *Kings Dethroned.* Miami: Hardpress Publishing, 1922 (original publication).

Hollander, Mark Steven. *Flat Truth*. Mark Steven Hollander, 2018.

Horgan, John. "Remembering Big Bang Basher Fred Hoyle." *Scientific American* website, August 25, 2008. https://www.scientificamerican.com/blog/cross-check/remembering-big-bang-basher-fred-hoyle/.

Hoskin, Michael. *The History of Astronomy: A Very Short Introduction.* Oxford: Oxford University Press, 2003.

Hougland, Richard C., and Mike Bara. *Dark Mission: The Secret History of NASA*. Port Townsend: Feral House, 2009.

Howell, Elizabeth. "Gus Grissom: 2nd American in Space." Space.com website, January 27, 2014. https://www.space.com/24443-gus-grissom.html.

Howell, Elizabeth, and Isabel Urrutia. "How Fast Is Earth Moving?" Space.com website, October 19, 2023. https://www.space.com/33527-how-fast-is-earth-moving.html.

Hoyle, Fred. *Nicolaus Copernicus: An Essay on His Life and Work.* London: Heinemann, 1973.

Hubble, Edwin. *The Observational Approach to Cosmology.* Oxford: Clarendon Press, 1937.

Hulsey, J. Leroy, and Zhili Quan. *A Structural Reevaluation of the Collapse of World Trade Center 7.* Prepared for Architects & Engineers for 9/11 Truth, March 2020. https://files.wtc7report.org/file/public-download/A-Structural-Reevaluation-of-the-Collapse-of-World-Trade-Center-7-March2020.pdf.

Humphreys, Russell. "Planetary Magnetic Dynamo Theories: A Century of Failure." The Proceedings on the International Conference of Creationism, 2013. https://digitalcommons.cedarville.edu/cgi/viewcontent.cgi?article=1212&context=icc_proceedings.

Hutchison, John. Affidavit. Wood v. Applied Research Associates, Inc., No. 07 CV 3314 (GBD) (S.D.N.Y.). https://www.thehutchisoneffect.com/docs/AffJHutchison4.pdf.

Hutchison, John, and ESJ staff. "The Hutchison Effect Apparatus." *Electric Spacecraft Journal*, Issue 9, 1993. https://www.thehutchisoneffect.com/docs/Hutchison%20Effect%20-%20More%20Info.pdf.

IAI TV. "The Dark Matter Myth – Pavel Kroupa." IAI TV website, n.d. https://iai.tv/video/the-dark-matter-myth-pavel-kroupa.

Idaho Statutes, "18-1506A Ritualized Abuse of a Child." The official website of the Idaho Legislature, n.d. https://legislature.idaho.gov/statutesrules/idstat/Title18/T18CH15/SECT18-1506A/.

"Interview with Brother Guy Consolmagno." *Astrobiology Magazine*, May 12, 2004. https://www.economicsvoodoo.com/wp-content/uploads/2004-05-12-Interview-with-Brother-Guy-Consolmagno-Astrobiology-Magazine.pdf.

"Interview with Max Planck." *The Observer*, January 25, 1931.

Jacobs, David. *Secret Life: Firsthand Documented Accounts of UFO Abductions.* New York: Fireside, 1992.

Jacobs, David. *Walking Among Us: The Alien Plan to Control Humanity*. San Francisco: Disinformation Books, 2015.

Jaffe, Bernard. *Michelson and the Speed of Light*. New York: Anchor Books, 1960.

Jahn, Robert, and Brenda Dunne. *Consciousness and the Source of Reality: The PEAR Odyssey*. Princeton, NJ: ICRL, 2011.

Jahn, Robert, and Brenda Dunne. *Margins of Reality: The Role of Consciousness in the Physical World*. Princeton, NJ: ICRL Press, 1981.

Jahn, Robert et al. "Correlations of Random Binary Sequences with Pre-Stated Operator Intention: A Review of a 12-Year Program." *Explore* 3, (2007): 244–53, 341–43. https://www.explorejournal.com/article/ S1550-8307(07)00062-6/fulltext.

James, Dave. *Escaping from Tyranny: True Earth and the Search for More Land Beyond the Antarctic Ice Ring*. Dave James, 2021.

James, Gary A., and Thomas D. Harris. "Calculation of Wind Compensation for Launching of Unguided Rockets." NASA website, April 17, 1961. https://ntrs.nasa.gov/api/citations/20040008097/downloads/20040008097.pdf.

Jarrell, Elizabeth M. "Robert Simmon – AKA Mr. Blue Marble." NASA website, June 12, 2012. https://www.nasa.gov/people-of-nasa/goddard-people/robert-simmon-aka-mr-blue-marble/.

Javanmardi, Behnam, and Pavel Kroupa. "Anisotropy in the All-Sky Distribution of Galaxy Morphological Types." *Astronomy & Astrophysics*, vol. 597, 2017, A120. https://arxiv.org/abs/1609.06719.

jeranism. "Long Range Shooting World Record Broken...What Was the Coriolis Correction?" YouTube video, November 7, 2022. https://www.youtube.com/watch?v=xPkCtAvNW2o.

Johnson Davis, Anne. *Hell Minus One: My Story of Deliverance from Satanic Ritual Abuse and My Journey to Freedom*. Tooele: Transcript Bulletin Publishing, Inc., 2008.

Kastrup, Bernardo. *Brief Peeks Beyond: Critical Essays on Metaphysics, Neuroscience, Free Will, Skepticism, and Culture*. Winchester, UK: Iff, 2015.

Kastrup, Bernardo. *The Idea of the World: A Multi-Disciplinary Argument for the Mental Nature of Reality*. Hampshire, UK: Iff, 2019.

Kastrup, Bernardo. *More Than Allegory: On Religious Myth, Truth and Belief.* Winchester, UK: John Hunt 2016.

Kastrup, Bernardo. "Transcending the Brain: At Least Some Cases of Physical Damage Are Associated with Enriched Consciousness or Cognitive Skill." *Scientific American* blog, March 29, 2017. https://blogs.scientificamerican. com/guest-blog/ transcending-the-brain/.

Kastrup, Bernardo. *Why Materialism Is Baloney: How True Skeptics Know There Is No Death and Fathom Answers to Life, the Universe and Everything*. Winchester, UK: Iff, 2014.

Katz, Jonathan I. The Biggest Bangs: *The Mystery of Gamma-Ray Bursts, The Most Violent Explosions in the Universe*. Oxford: Oxford University Press, 2002.

Kaufman, Andrew, M.D. "Fallacies, Fraud & Pseudoscience: The Parallels Between Cosmology and Biology." Rumble video, January 2024. https://rumble.com/v44kwso-fallacies-fraud-and-pseudoscience-the-parallels-between-cosmology-and-biolo.html.

Keefe, Josh, and Matthew Beavers. *Truth About Cosmology: Overcoming Doctrines of Demons*. Matthew Beavers and Josh Keefe, 2022.

Kimball, Daryl. "The Outer Space Treaty at a Glance." Arms Control Association website, 2024. https://www.armscontrol.org/factsheets/outer-space-treaty-glance.

Kinsella, Philip. *Terrestrial Trespassers: The Greys, Abductions & Areas of High Strangeness*. Pontefract, Flying Disk Press, 2023.

Kinzer, Stephen. *Poisoner in Chief: Sidney Gottlieb and the CIA Search for Mind Control.* New York: Henry Holt and Company, 2019.

Kositzke, Richard. "Secret Space Program & Flat Earth Dome /w Aug Tellez [Strongholds]." YouTube video, September 8, 2016. https://www.youtube.com/watch?v=kH6wFf0eMWU.

Kostro, Ludwik. *Einstein and the Ether*. Montreal: Apeiron, 2000.

Kranz, Gene. *Failure Is Not an Option*. New York: Simon & Schuster Paperbacks, 2000.

Kroupa, Pavel. "The Dark Matter Crisis: Falsification of the Current Standard Model of Cosmology." *Publications of the Astronomical Society of Australia*, 29 (2012): 395-433. https://www.cambridge.org/core/services/aop-cambridge-core/

content/view/E3DB55A34BB36EA723671D1F54905B30/S1323358000001417a.pdf/dark_matter_crisis_falsification_of_the_current_standard_model_of_cosmology.pdf.

Kroupa, Pavel, and Michael Haslbauer. "Our Model of the Universe Has Been Falsified." IAI News website, February 14, 2023. https://iai.tv/articles/our-model-of-the-universe-has-been-falsified-auid-2393.

Kuhn, Thomas. *The Copernican Revolution: Planetary Astronomy in the Development of Western Thought.* Cambridge: Harvard University Press, 1957.

Kuhn, Thomas. *The Structure of Scientific Revolutions.* Chicago: University of Chicago Press, 1970.

Larson, Cynthia Sue. *The Mandela Effect and Its Society: Awakening from ME to WE.* Independently published, 2024.

Larson, Dewey. *The Case Against the Nuclear Atom.* Portland: North Pacific Publishers, 1963.

Lee, James. *Flat Earth: Investigations Into a Massive 500-Year Heliocentric Lie.* CreateSpace Independent Publishing Platform, 2017.

Lee, James W. *The One World Tartarians: The Greatest Civilization Ever to Be Erased from History.* Independently published, 2020.

Leeman, John. "The Deepest Borehole on Earth – Don't Panic Podcast, Episode 350." Leeman Geophysical website, December 4, 2022. https://leemangeophysical.com/the-deepest-borehole-on-earth/.

Lester, Dawn, and David Parker. *What Really Makes You Ill? Why Everything You Thought You Knew about Disease Is Wrong.* Dawn Lester and David Parker, 2019.

Level Earth Observer. "Flat Earth: I Love Destroying the Globe." YouTube video, April 21, 2022. https://www.youtube.com/watch?v=R5PPaJkmpDc.

Liedtke, Martin. *The Holy Grail of Our Flat Earth.* Martin Liedtke, 2019.

Liles, Maryn. "50 Mandela Effect Examples of Things You *Think* You Remember Correctly (That You've Actually Got All Wrong)." *Parade* website, August 2, 2023. https://parade.com/1054775/marynliles/mandela-effect-examples/.

Liverman v. McMurray, US Supreme Court, No. 20-5808. "In re. Roger Liverman, PRO SE PETITIONER vs. Lawanda McMurray, et al, RESPONDENT(S) ON PETITION FOR A WRIT OF MANDAMUS, TO THE COURT OF APPEALS, FIFTH CIRCUIT." Filed August 26, 2020, docketed September 25, 2020. https://www.supremecourt.gov/DocketPDF/20/20-5808/176392/20210422160651080_20210422-160534-95753440-00001842.pdf.

Long, Matt. *The House That Jesus Built*. Carbfort Publishing, 2021.

Lucas, Anneke. *Quest for Love: Memoir of a Child Sex Slave*. Unconditional Books, 2022.

Lyne, William. *Occult Ether Physics: Tesla's "Ideal Flying Machine" and the Conspiracy to Conceal It*. New Mexico: CREATOPIA, 1997.

Mack, John. *Abduction: Human Encounters with Aliens*. New York: Macmillan Publishing Company, 1994.

Mack, John E. "The Believer: Ralph Blumenthal and Leslie Kean on Alien Encounters." YouTube video, May 23, 2021. https://www.youtube.com/watch?v=BMOBoaSqegc.

Mack, John. *Passport to the Cosmos: Human Transformation and Alien Encounters*. Guildford, UK: White Crow Books, 2011.

Malashenko, Uliana. "Fact Check: Polish Astronaut Did NOT Admit The Earth Is Flat—It Was a Joke." Lead Stories website, July 13, 2023. https://leadstories.com/hoax-alert/2023/07/fact-check-polish-astronaut-did-not-admit-the-earth-is-flat.html.

Mallan, Lloyd. *Russia's Space Hoax: Documented Proof That the Soviet Space Program Has Been Faked, Over 14 Months of Research Uncovers Evidence from 58 Authorities*. Science & Mechanics, News Book, 1966.

Mark, Joshua J. "Greek Astronomy." *World History Encyclopedia* website, n.d. https://www.worldhistory.org/Greek_Astronomy/.

Markolin, Caroline. "THE FIVE BIOLOGICAL LAWS OF THE NEW MEDICINE." Learning GNM website, n.d. https://learninggnm.com/SBS/documents/five_laws.html.

Masayuki, Ohkado. *Katsugoro and Other Reincarnation Cases in Japan*. Santa Fe: Afterworlds Press, 2024.

Mayweather, Kelvin. *Flat Earth Debunked: The Simple Science Guide to Why the Earth Is a Globe*. Flow Publishing88, 2023.

Mazzucco, Massimo. *American Moon*. Luogocomune, 2017. https://www.luogocomune.net/shop/films-in-english/american-moon-english-version.

McDougall, Walter A. ... *The Heavens and Earth: A Political History of the Space Age*. Baltimore: The Johns Hopkins University Press, 1985.

McFall-Johnsen, Morgan. "Earth is screaming through space at 1.3 million mph. A simple animation by a former NASA scientist shows what that looks like." *Business Insider* website, October 11, 2019. https://www.businessinsider.com/earth-screaming-through-space-nasa-animated-video-2019-10.

McRae, Mike. "Physicists Find No Sign of the Particle That Made the Universe Explode." *ScienceAlert* website, June 10, 2017. https://www.sciencealert.com/physicists-find-no-sign-of-the-particle-that-made-the-universe-explode.

"Memorandum for the Secretary of Defense, Unclassified." March 13, 1962. https://archive.org/details/OperationNorthwoods.

Merali, Zeeya. "'Axis of evil' a cause for cosmic concern." *New Scientist* website, April 3, 2007. https://www.newscientist.com/article/mg19425994-000-axis-of-evil-a-cause-for-cosmic-concern/.

Meyer, Marvin (Ed.). *The Nag Hammadi Scriptures*. New York: Harper One, 2007.

Meyer, Stephen C. *Return of the God Hypothesis: Three Scientific Discoveries That Reveal the Mind Behind the Universe*. New York: HarperOne, 2020.

Michels, Patrick. "Granite and Grace at Llano's World Championship of Rock Stacking." *Texas Monthly* website, April 5, 2023. https://www.texasmonthly.com/being-texan/llano-world-rock-stacking-championship/.

Michelson, Albert A. "The Relative Motion of the Earth and the Luminiferous Ether." *American Journal of Science*, vol. 22, no. 128, August 1881, pp. 120–29.

Miletta, Joseph. "Propagation of Electromagnetic Fields Over Flat Earth." Army Research Laboratory, February 2001. https://ia903206.us.archive.org/23/items/propagation_of_electromagnetic_fields_over_flat_earth/propagation_of_electromagnetic_fields_over_flat_earth.pdf.

Miller, Arthur I. *Albert Einstein's Special Theory of Relativity: Emergence (1905) and Early Interpretation (1905–1911)*. Reading, Mass.: Addison-Wesley, 1981.

Mitchell fromAustralia. "At Best, All YOU'VE Done, Is Debunk the Rotation of the Earth!" YouTube video, August 9, 2022. https://www.youtube.com/watch?v=sQNWMuqqsFQ.

Mitchell fromAustralia. "FLAT EARTH SCHOOL – Why Objects Disappear Bottom Up." YouTube video, December 26, 2021. https://www.youtube.com/watch?v=aVVbsekJ9Sg.

MoneyBags73. "Mandela Effect Passenger Airliners Look Different In This Reality!! Voting Video #248." YouTube video, April 28, 2018. https://www.youtube.com/watch?v=IBw1CkU5Xeo.

Moore, Cherice, Randall Svetlik, and Antony Williams. "Practical Applications of Cables and Ropes in the ISS Countermeasures System." Presented at the 2017 IEEE Aerospace Conference, Big Sky, MT, USA, March 4–11, 2017. Published by Institute of Electrical and Electronics Engineers (IEEE). https://ntrs.nasa.gov/api/citations/20160012382/downloads/20160012382.pdf.

Moorjani, Anita. "The Day My Life Changed: February 2, 2006." Anita Moorjani webpage, n.d. http://anitamoorjani.com/about-anita/near-death-experience-description/.

Moorjani, Anita. *Dying to Be Me: My Journey from Cancer, to Near Death, to True Healing*. Carlsbad, CA: Hay House, 2012.

Morgan, Richard Jameson. *The Iron Republic.* Point Pleasant: Heathen Editions, 2023.

Moskowitz, Clara. "What's 96 Percent of the Universe Made Of? Astronomers Don't Know." Space.com, May 12, 2011. https://www.space.com/11642-dark-matter-dark-energy-4-percent-universe-panek.html.

Mossbridge J., P. Tressoldi, and J. Utts. "Predictive Physiological Anticipation Preceding Seemingly Unpredictable Stimuli: A Meta-Analysis." *Frontiers in Psychology*, 3 (2012): 390. https://www.frontiersin.org/articles/10.3389/ fpsyg.2012.00390/full.

Moy, John. "Science, Religion, and the Galileo Affair." Center for Inquiry website, September 2001. https://cdn.centerforinquiry.org/wp-content/uploads/sites/29/2001/09/22164801/p43.pdf.

MythologyWorldwide. "The Myth of the Ten Suns in Chinese Mythology." MythologyWorldwide website, February 29, 2024. https://mythologyworldwide.com/the-myth-of-the-ten-suns-in-chinese-mythology/.

Na, Sung-Ho et al. "Chandler Wobble and Free Core Nutation: Theory and Features." *Journal of Astronomy and Space Sciences*, vol. 36, 2019, pp. 11–18. https://adsabs.harvard.edu/pdf/2019JASS...36...11N#:~:text=Chandler%20wobble%20is%20a%20prograde,latter%20is%20free%20core%20nutation.

Naboulsi, Marouane, S. Haddad, and M. Sizun. "Fog attenuation prediction for optical and infrared waves." *Optical Engineering*, vol. 43, no. 2, 2004, pp. 319–29. https://proceedings.spiedigitallibrary.org/journals/optical-engineering/volume-43/issue-2/0000/Fog-attenuation-prediction-for-optical-and-infrared-waves/10.1117/1.1637611.short.

NASA. "Apollo 8 Mission Report." Manned Spacecraft Center, February 1969. https://www.nasa.gov/wp-content/uploads/static/history/alsj/a410/A08_MissionReport.pdf.

NASA. "The Apollo 11 Telemetry Data Recordings: A Final Report." NASA, January 1, 2010. https://www.nasa.gov/wp-content/uploads/static/history/alsj/a11/apollo_11_tv_tapes_report.pdf.

NASA. "Apollo 14 Mission Report." Manned Spacecraft Center, May 1971. https://www.nasa.gov/wp-content/uploads/static/history/alsj/a14/A14MRntrs.pdf.

NASA. "Edwin Hubble." NASA website, n.d. https://science.nasa.gov/people/edwin-hubble/.

NASA. "Marshall Space Flight Center Electrostatic Levitation Laboratory." NASA website, n.d. https://www.nasa.gov/wp-content/uploads/2016/01/g-29859_esl.pdf.

NASA. "The NACA." NASA website, n.d. https://www.nasa.gov/history/the-national-advisory-committee-for-aeronautics-naca/.

NASA. "The NASA MSFC Electrostatic Levitation (ESL) Laboratory: Summary of Capabilities, Recent Upgrades, and Future Work." NASA website, n.d. https://ntrs.nasa.gov/citations/20150022349.

NASA. "Training for Apollo 9." NASA website, June 9, 2009. https://www.nasa.gov/image-article/training-apollo-9/.

NASA. "What Is the Inflation Theory?" NASA website, n.d. https://wmap.gsfc.nasa.gov/universe/bb_cosmo_infl.html

NASA's Earth Observatory. "The Blue Marble." NASA website, October 13, 2005. https://earthobservatory.nasa.gov/features/BlueMarble/BlueMarble_2002.php.

NASA Earth Observatory. "Fisheye Over Sinai." NASA Earth Observatory website, August 18, 2019. https://earthobservatory.nasa.gov/images/147993/fisheye-over-sinai.

NASA Earth Observatory. "Planetary Motion: The History of an Idea That Launched the Scientific Revolution." NASA website, n.d. https://earthobservatory.nasa.gov/features/OrbitsHistory/.

NASA Hubblesite. "NASA's Webb, Hubble Telescopes Affirm Universe's Expansion Rate, Puzzle Persists." NASA website, n.d. https://hubblesite.org/contents/news-releases/2024/news-2024-108.html.

NASA. "Neutral Buoyancy Laboratory." NASA website, n.d. https://www.nasa.gov/johnson/neutral-buoyancy-laboratory/.

NASA. "Nine Ways We Use AR and VR on the International Space Station." NASA website, September 20, 2021. https://www.nasa.gov/missions/station/nine-ways-we-use-ar-and-vr-on-the-international-space-station/.

NASA Photojournal. "PIA18176: Galileo Over Io (Artist's Concept)." NASA Photojournal website, n.d. https://photojournal.jpl.nasa.gov/mission/Galileo.

NASA. "Scientific Balloons." NASA website, n.d. https://www.nasa.gov/scientificballoons/.

NASA Science Editorial Team. "Galileo's Observations of the Moon, Jupiter, Venus and the Sun." NASA website, n.d. https://science.nasa.gov/solar-system/galileos-observations-of-the-moon-jupiter-venus-and-the-sun/.

NASA Science Space Place. "What Is Dark Matter?" NASA Space Place website, n.d. https://spaceplace.nasa.gov/dark-matter/en/.

NASA. "Tests of Big Bang: The CMB." NASA website, n.d. https://map.gsfc.nasa.gov/universe/bb_tests_cmb.html.

National Archives. "JFK Assassination Records: Summary of Findings," based on 1979 report. https://www.archives.gov/research/jfk/select-committee-report/summary.html.

National Geographic. "Rotation." *National Geographic* website, n.d. https://education.nationalgeographic.org/resource/rotation/.

Nedic, Vladimir. "Thermo-Gravitational Oscillator and Tesla's Energy Source Ether." *International Tesla Congress 2015, The History of the Future*, April 24–25, 2015. https://www.researchgate.net/publication/280304207_Thermo-Gravitational_Oscillator_and_Tesla's_Energy_Source_Ether.

New Thinking Allowed Foundation. "The Mandela Effect with Cynthia Sue Larson." New Thinking Allowed Foundation website, n.d. https://www.newthinkingallowed.org/the-mandela-effect-with-cynthia-sue-larson/.

Newton, Isaac. *The Principia*. Berkeley: University of California Press, 1999.

Niiler, Eric. "Hitler Sent a Secret Expedition to Antarctica in a Hunt for Margarine Fat." *History* website, July 27, 2018. https://www.history.com/news/hitler-nazi-secret-expedition-antarctica-whale-oil.

The 92nd Street Y, New York. "Neil deGrasse Tyson Explains How the Earth Became Pear-Shaped." YouTube video, February 10, 2014. https://www.youtube.com/watch?v=SoCKapivHGM.

Nordregio. "Protected areas in the Arctic." Nordregio website, January 2019. https://nordregio.org/maps/protected-areas-in-the-arctic/.

Norling, Ernest R. *Perspective Made Easy*. Garden City: Dover Publications: 1999.

Nosachev, Pavel. "The Influences of Western Esotericism on Russian Rock Poetry of the Turn of the Century." In *Esotericism, Literature and Culture in Central and Eastern Europe*, edited by Nemanja Radulović, 201–20. Belgrade, 2018. https://publications.hse.ru/pubs/share/direct/220857556.pdf.

NOVA PBS Official. "Arctic Sinkholes I Full Documentary I NOVA I PBS." YouTube video, February 2, 2022. https://www.youtube.com/watch?v=HvKpnaXYUPU.

NPR. "Iraq WMD Timeline: How the Mystery Unraveled." NPR website, November 15, 2005. https://www.npr.org/2005/11/15/4996218/iraq-wmd-timeline-how-the-mystery-unraveled.

Odle, Dean. *Like Clay Under the Seal*. Dean Odle, 2019.

Old World Order. 2024, https://owofilm.com/.

O'Neill, Tom, with Dan Piepenbring. *CHAOS: Charles Manson, the CIA, and the Secret History of the Sixties*. New York: Hachette Book Group, 2019.

Online Etymology Dictionary. "Planet (n.)." Online Etymology Dictionary website, n.d. https://www.etymonline.com/word/planet.

"Out of Shadows: The Documentary." https://www.outofshadows.org/.

Owocki, Stan. *Fundamentals of Astrophysics*. Cambridge: Cambridge University Press, 2021.

Papadopulos-Eleopulos, E. et al. "HIV—A virus like no other." Perth Group website, July 12, 2017. http://www.theperthgroup.com/HIV/TPGVirusLikeNoOther.pdf.

Paterson, Pat. "The Truth about Tonkin." U.S. Naval Institute website, February 2008, https://www.usni.org/magazines/naval-history-magazine/2008/february/truth-about-tonkin.

Paul Davis UnCancelled. "Are the astronauts from the Challenger disaster still alive? Justin Harvey has the evidence." Rumble video, June 20, 2024. https://rumble.com/v52mvyw-are-the-astronauts-from-the-challenger-disaster-still-alive-justin-harvey-h.html.

Paul On The Plane. "Antarctica Is NOT a Continent | Tiger Dan925 Jumped Ship (Jeranism Mirror)." YouTube video, April 6, 2018. https://www.youtube.com/watch?v=YLkTm19aZGs.

Pauli, Wolfgang. *Theory of Relativity*. New York: Pergamon Press, 1958.

PBS NOVA. "A Tainted Legacy." PBS website, n.d. https://www.pbs.org/wgbh/nova/sputnik/vonb-02.html.

Pearlman, Robert Z. "How did CBS' Big Bang Theory send Howard Wolowitz to space?" Space.com, March 15, 2022. https://www.space.com/17891-big-bang-theory-space-set-design.html.

Pearsall, Paul. *The Heart's Code: Tapping the Wisdom and Power of Our Heart Energy; The New Findings About Cellular Memories and Their Role in the Mind, Body, Spirit Connection*. New York: Broadway, 1999.

Pettit, Harry. "Panicked NASA cancels spacewalks after ISS astronaut's helmet 'fills with water.'" *New York Post*, May 19, 2022.

https://nypost.com/2022/05/19/panicked-nasa-cancels-spacewalks-after-iss-astronauts-helmet-fills-with-water/.

Planet Plane. "Eric Dubay: The History of Flat Earth." YouTube video, January 20, 2017. https://www.youtube.com/watch?v=9496sa0cXh4.

Poincaré, Henri. *La Science et l'Hypothèse* ("Science and Hypothesis"). Paris: Flammarion, 1901. Reprint, 1968.

Poincaré, Henri. "The Principles of Mathematical Physics." *The Monist*, vol. XV, January 1905. http://www.jstor.org/stable/27899559?seq=20.

Poor, Charles Lane. *Gravitation versus Relativity*. New York: G. P. Putnam's Sons, 1922.

Popov, Luka. "Newtonian–Machian Analysis of the Neo-Tychonian Model of Planetary Motions." *European Journal of Physics*, vol. 34, 2013, pp. 383–91.

Prados, John. "Essay: 40th Anniversary of the Gulf of Tonkin Incident." *National Security Archive*, August 4, 2004. https://nsarchive2.gwu.edu/NSAEBB/NSAEBB132/essay.htm.

Prasad, Ravi. Instagram post, June 24, 2024. https://www.instagram.com/reel/C8msOW4SbAB/?igsh=MXV4NGNoMTJ5OXhwbw%3D%3D.

Priest, Aaron, and Michael J. Pechan. "The Driving Mechanism for a Foucault Pendulum (Revisited)." *American Journal of Physics* 76, 188 (2008). https://pubs.aip.org/aapt/ajp/article-abstract/76/2/188/1057096/The-driving-mechanism-for-a-Foucault-pendulum?redirectedFrom=fulltext.

Project Petri. "Admiral Richard E. Byrd 1954 Interview, Enhanced Audio." YouTube video, June 16, 2023. https://www.youtube.com/watch?v=PsheO3qh2dw.

Ptolemy, Claudius. *The Almagest*. Santa Fe: Green Lion Press, 2021.

Question Everything. "Heliosorcery." Rumble video, 2022. https://rumble.com/v4ybjcq-heliosorcery-2022-exposing-the-occult-origins-of-heliocentrism-full-documen.html.

Radin, Dean. *Entangled Minds Extrasensory Experiences in a Quantum Reality*. New York: Paraview, 2006.

Radin, Dean. *Real Magic: Ancient Wisdom, Modern Science, and a Guide to the Secret Power of the Universe*. New York: Harmony, 2018.

Radin, Dean. *Supernormal: Science, Yoga, and the Evidence for Extraordinary Psychic Abilities*. New York: Random House, 2013.

Radin, Dean, Nelson Roger, Dobyns York, and Houtkooper Joop. "Reexamining psychokinesis: comment on Bösch, Steinkamp, and Boller" (2006). *Psychological Bulletin*. 2006;132(4):529–32; discussion 533–7.

Rafelski, Johann. "Critical Acceleration and Vacuum Structure." Paper presented at the Charles A. Whitten Memorial Symposium, Department of Physics, the University of Arizona, December 15–16, 2011. http://www.physics.arizona.edu/~rafelski/PS/1112WhittenUCLA.pdf.

RAND. "U.S. Armed Forces Capabilities in Arctic Region Pose National Security Risks, Need Strengthening." RAND website, November 1, 2023. https://www.rand.org/news/press/2023/11/01.html.

REALWOMEN/REALSTORIES, "(UNCUT) Anneke Lucas: Sold by my mother into a pedophile network at six years of age." YouTube video, November 26, 2020. https://www.youtube.com/watch?v=RTxD6VUcxZ0.

Reville, William. "Remarkable story of maths genius who had almost no brain." *The Irish Times*, November 9, 2006. https://www.irishtimes.com/news/remarkable-story-of-maths-genius-who-had-almost-no-brain-1.1026845.

Rieke, George H. "Ptolemy." University of Arizona website, n.d. https://ircamera.as.arizona.edu/NatSci102/NatSci102/lectures/ptolemy.htm.

Rivas, Titas, Anny Dirven, and Rudolf H. Smit. *The Self Does Not Die: Verified Paranormal Phenomena from Near-Death Experiences*. Durham, NC: IANDS Publications, 2016 (second edition, 2023).

Roberts, Nathan. *The Doctrine of the Shape of the Earth*. Nathan Roberts, 2017.

RobinsHoodlum. "Globe Earth SNIPER ~ from Mountainbear." Rumble video, August 27, 2022. https://rumble.com/v1hlc50-globe-earth-sniper-from-mountainbear.html.

Rosenberg, Matt. "Speed of the Earth." ThoughtCo website, February 13, 2018. https://www.thoughtco.com/speed-of-the-earth-1435093.

Rothman, Aviva (Ed.). *The Dawn of Modern Cosmology: From Copernicus to Newton*. Dublin: Penguin Classics, 2023.

Rovelli, Carlo. *Anaximander*. New York: Westholme Publishing, 2011.

Rowbotham, Samuel Birley. *Zetetic Astronomy: Earth Not a Globe*. Pantianos Classics, 1881.

RunBostonBear. "BRAVERITAS with Austin Whitsitt (Day 1) • Live at Old State Saloon on 06/04/24." YouTube video, June 6, 2024. https://www.youtube.com/watch?v=ssCZXJrDxiQ.

Russell, Peter. *From Science to God: The Mystery of Consciousness and the Meaning of Light*. Novato: New World Library, 2003.

Salla, Michael. *Antarctica's Hidden History*. Pahoa: Michael E. Salla, PhD, 2018.

Samuelson, Kate. "Take a Tour of the International Space Station In Dazzling 4K Detail." *Time*, October 28, 2016. https://time.com/4549095/nasa-international-space-station-4k/.

Schrödinger, Erwin. *What Is Life? With Mind and Matter and Autobiographical Sketches.* London: Cambridge University Press, 1969.

Schumm, Laura. "What Was Operation Paperclip?" History.com, March 4, 2020. https://www.history.com/news/what-was-operation-paperclip.

Schwarz, Dominik J. et al. "Is the Low-ℓ Microwave Background Cosmic?" *Physical Review Letters*, vol. 93, 2004, pp. 221–28.

Science|Business. "Conspiracy Thinking: 25% of Europeans Say the Moon Landing Never Happened." *Science|Business* website, January 25, 2022. https://sciencebusiness.net/news/covid-19/conspiracy-thinking-25-europeans-say-moon-landing-never-happened.

Science History Institute. "The Sex-Cult 'Antichrist' Who Rocketed Us to Space." Science History Institute website, n.d. https://www.sciencehistory.org/stories/disappearing-pod/the-sex-cult-antichrist-who-rocketed-us-to-space-part-1/.

Scoles, Sarah. "Will Scientists Ever Find a Theory of Everything?" *Scientific American* website, August 19, 2023. https://www.

scientificamerican.com/article/will-scientists-ever-find-a-theory-of-everything/.

Scott, David Wardlaw. *Terra Firma: The Earth Is Not a Planet*. London: Simpkin, Marshall, Hamilton, Kent & Co., Ltd.: 1901.

Shackleton, Sir Ernest. *South: Shackleton's Last Expedition 1914–1917*. Orinda: Sea Wolf Press, 2020.

Sharma, Poonam, and Jim Tucker, MD. "Cases of the Reincarnation Type with Memories from the Intermission Between Lives." *Journal of Near-Death Studies*, 23 (2) (2004), 101, 118.

Sibrel, Bart. "Moon Landing Fraud – Deathbed Confession – MM16." YouTube video, September 13, 2022. https://www.youtube.com/watch?v=wu5Z75ji3aU.

Sibrel, Bart. *Moon Man: The True Story of a Filmmaker on the CIA Hit List*. Washington, DC: Bart Sibrel, 2021.

Sibrel, Bart. "NASA Admits Fraud – MM10 – READ BELOW – Links 10 B-C-D." YouTube video, January 11, 2019, 6:30–7:10. https://www.youtube.com/watch?v=x6WAuXjQDoo.

60 Minutes Australia (1989). "Satan's Children: Satanic Ritual Abuse." Bitchute video, January 26, 2020. https://www.bitchute.com/video/I1bHD2JyfmEg/.

Skeptiko. "Dr. Mario Beauregard, Frontier Science Wake Up Call |538|." YouTube video, February 1, 2022. https://www.youtube.com/watch?v=8i-Q0MTJFls&t=2114s.

Skeptiko. "Mike Clelland, Owls and Extended Consciousness |387|." YouTube video, August 21, 2018. https://www.youtube.com/watch?v=GQFGFccl5Gc.

Skiba, Rob. "The Chicago Skyline Expanded Edition – Part 1: Refraction, Magnification or Curvature?" YouTube video, January 27, 2017. https://www.youtube.com/watch?v=I-Q-FuXJSTQ&t=1s.

Skiba, Rob (Ed.). *Testing the Globe: A Zetetic Investigation Volume I*. The Colony, King's Gate Media LLC, 2018.

Smith, Ashea. "Travis and Jason Kelce Reveal How Many NFL Players Believe in 'Flat Earth' Conspiracy Theory: 'You Would Be Shocked.'" *MSN* website, May 8, 2024. https://www.msn.com/en-us/sports/nfl/travis-and-jason-kelce-reveal-how-many-nfl-players-believe-in-flat-earth-conspiracy-theory-you-would-be-shocked/ar-BB1m2K92.

Smithsonian National Air and Space Museum. "Lens, 16mm, Fisheye, Nikon, Shuttle." Smithsonian National Air and Space Museum website, n.d. https://airandspace.si.edu/collection-objects/lens-16mm-fisheye-nikon-shuttle/nasm_A20080106000.

Snow, Donna. "The Mandela Effect and the Bible." Artesian Ministries website, January 12, 2023. https://www.artesianministries.org/faith/the-mandela-effect-and-the-bible/.

Snow, John B. "New World Record Set for Farthest Long-Range Rifle Shot: 4.4 Miles." *Outdoor Life* website, September 20, 2022. https://www.outdoorlife.com/guns/new-world-record-longest-rifle-shot/.

Sokol, Joshua. "How the Ancient Art of Eclipse Prediction Became an Exact Science." *Quanta Magazine* website, April 5, 2024. https://www.quantamagazine.org/how-the-ancient-art-of-eclipse-prediction-became-an-exact-science-20240405/.

The Sol Foundation. "Iya Whiteley, Ph.D. on Trusting and Learning from Pilots." YouTube video, February 12, 2024. https://www.youtube.com/watch?app=desktop&v=HR09GHQ5AwA.

Spectator staff. "Is the Earth Round or Flat? Culture Ministry Chief Official Says Proof Is Still Needed." *The Slovak Spectator* website, May 3, 2024. https://spectator.sme.sk/c/23325477/is-the-earth-round-or-flat-culture-ministry-chief-official-says-proof-is-still-needed.html?ref=njctse.

Spivack, Barry, and Patricia Anne Saunders. *An Antidote to Violence.* Winchester: Changemakers Books, 2020.

Stanford Encyclopedia of Philosophy. "Nicolaus Copernicus." *Stanford Encyclopedia of Philosophy* website, n.d. https://plato.stanford.edu/entries/copernicus/.

Steele, Robert David. "RAW: Sean Stone with Jay Parker," Robert David Steele website. March 2, 2021, https://robertdavidsteele.com/sean-stone-with-jay-parker/.

Stellium7 YouTube Channel. https://www.youtube.com/@stellium7.

Stevenson, Ian. *Children Who Remember Previous Lives: A Question of Reincarnation.* Jefferson, NC: McFarland, 2001.

Stevenson, Ian. *Where Reincarnation and Biology Intersect.* Westport, CT: Praeger, 1997.

Stinnett, Robert B. *Day of Deceit: The Truth about FDR and Pearl Harbor*. New York: Touchstone, 2001.

Strassman, Rick. *DMT: The Spirit Molecule: A Doctor's Revolutionary Research into the Biology of Near-Death and Mystical Experiences.* Rochester, NY: Park Street Press, 2001.

Sullivan, Casey C. "75-Year-Old Woman, Detained by NASA for Selling a Moon Rock, Can Sue." *FindLaw* website, March 21, 2019. https://www.findlaw.com/legalblogs/ninth-circuit/75-year-old-woman-detained-by-nasa-for-selling-a-moon-rock-can-sue/.

Sunfellow, David. "Near-Death Experiences & Miraculous Healings." David Sunfellow's Substack website, April 16, 2023. https://sunfellow.substack.com/p/near-death-experiences-and-miraculous.

Sungenis, Robert. *Flat Earth/Flat Wrong: An Historical, Biblical & Scientific Analysis.* 2018.

Sungenis, Robert. *Geocentrism 101: An Introduction into the Science of Geocentric Cosmology*. United States: CAI Publishing, Inc., 2013.

Sungenis, Robert. "The Principle." YouTube video, October 18, 2022. https://www.youtube.com/watch?v=CeJb0JIHNik.

Swartz, Tim R. *Admiral Byrd's Secret Journey Beyond the Poles*. New Brunswick: Global Communications, 2011.

Swedenborg, Emmanuel. *Other Planets*. West Chester, PA: Swedenborg Foundation, 2018.

Taboo Conspiracy. "How Time Zones Hide the Flat Earth by Hervé Riboni." YouTube video, January 28, 2023. https://www.youtube.com/watch?v=lMhheDWThxE.

Taboo Conspiracy. "Sorry, Antarctica Is Closed." YouTube video, November 10, 2020. https://www.youtube.com/watch?v=SmYRFtY_jfQ.

Taboo Conspiracy. "USSR Cosmonauts: Earth Is Flat and Nobody Has Been to Space (Old Taboo)." YouTube video, May 9, 2023. https://www.youtube.com/watch?v=2oUzONrecXc.

TEDx Talks. "From Child Sex Slavery to Victory — My Healing Journey | Anneke Lucas | TEDxKlagenfurt." YouTube video, June 26, 2017. https://www.youtube.com/watch?v=tLp5LM18wIU

The Telegraph. "Moon Rock Given to Holland by Neil Armstrong and Buzz Aldrin Is Fake." *The Telegraph* website, August 29, 2009. https://www.telegraph.co.uk/news/science/space/6105902/Moon-rock-given-to-Holland-by-Neil-Armstrong-and-Buzz-Aldrin-is-fake.html.

Tellinger, Michael. "A Time Before the Moon Was in the Sky." Michael Tellinger website, February 20, 2018. https://www.michaeltellinger.com/a-time-before-the-moon-was-in-the-sky/.

Tesla, Nikola. "Radio Power Will Revolutionize the World." Tesla Universe website, July 1934 article. https://teslauniverse.com/nikola-tesla/articles/radio-power-will-revolutionize-world.

thisisflatearth. "Don Pettit – I'd Go to the Moon in a Nanosecond [13]." Rumble video, January 1, 2024. https://rumble.com/v44jwbc-don-pettit-id-go-to-the-moon-in-a-nanosecond-.html.

Thompson v. Garcia, Case No. 2019-MV-1104 (Magistrate Court of Barrow County, 2019). https://rodscontracts.com/docs/legal/ThompsonVersusGarcia.pdf.

Tilak, Bal Gangadhar. *The Arctic Home in the Vedas*. London: Arktos, 2011.

Tsakiris, Alex. "Anneke Lucas, Recovering From Unimaginable Evil |441|." Skeptiko website, February 18, 2020, https://skeptiko.com/anneke-lucas-recovering-from-unimaginable-evil-441/.

Tunick, Arnold. "An Energy Budget Model to Calculate the Low Atmosphere Profiles of Effective Sound Speed at Night." Army Research Laboratory, May 2003. https://apps.dtic.mil/sti/pdfs/ADA414792.pdf.

Tyson, Neil deGrasse. *Astrophysics for People in a Hurry*. New York: W.W. Norton & Company, 2017.

University of Maryland Department of Physics. "Lathrop Lab's Geodynamo Set for Overhaul." University of Maryland website, March 6, 2020. https://umdphysics.umd.edu/about-us/news/research-news/1562-lathrop-lab-s-geodynamo-set-for-overhaul.html.

Urban, Hugh B. *The Church of Scientology: A History of a New Religion*. Princeton: Princeton University Press, 2011.

Uri, John. "410 Years Ago: Galileo Discovers Jupiter's Moons." NASA website, January 9, 2020. https://www.nasa.gov/history/410-years-ago-galileo-discovers-jupiters-moons/.

US Department of Homeland Security. "Kona Blue" (presentation). AARO website, February 5, 2024. https://www.aaro.mil/Portals/136/PDFs/UAP_RECORDS_RESEARCH/AARO_DHS_Kona_Blue.pdf?ver=BjOpTzFISPc0LWMw5uAzzw%3d%3d.

Utts, Jessica. "An Assessment of the Evidence for Psychic Functioning." *Journal of Scientific Exploration* 10, no. 1 (1996): 3–30.

Vallee, Jacques, and Chris Aubeck. *Wonders in the Sky: Unexplained Aerial Objects from Antiquity to Modern Times*. New York: Penguin Group, 2010.

Vallee, Jacques, and Eric W. Davis. "Incommensurability, Orthodoxy and the Physics of High Strangeness: A 6-layer Model for Anomalous Phenomena." October 24, 2003. https://www.jacquesvallee.net/wp-content/uploads/2018/11/Incommensurability_Orthodoxy_and_the_Phy.pdf.

Vallee, Jacques. *The Invisible College: What a Group of Scientists Has Discovered about UFO Influence on the Human Race*. San Antonio, TX: Anomalist Books, 2014.

Vallee, Jacques. *Passport to Magonia: From Folklore to Flying Saucers*. Brisbane: Daily Grail Publishing, 2014

Vallee, Jacques. *Revelations: Alien Contact and Human Deception*. San Antonio: Anomalist Books, 1991.

Varshni, Y. P. "The Red Shift Hypothesis for Quasars: Is the Earth the Center of the Universe?" *Astrophysics and Space Science*, vol. 43, 1976, pp. 3–8.

Vincent, Alice. "Sex, rocket science, and Scientology: Meet NASA's real hidden figures." *The Telegraph*, February 17, 2017. https://www.telegraph.co.uk/films/2017/02/17/sex-rocket-science-scientology-meet-hidden-figures-behind-nasas/.

Virginia Institute of Marine Science. "The Equilibrium Theory of Tides." Virginia Institute of Marine Science website, n.d. https://www.vims.edu/research/units/labgroups/tc_tutorial/static.php.

ViroLIEgy website. https://viroliegy.com/.

Wahbeh, Helané. *The Science of Channeling: Why You Should Trust Your Intuition & Embrace the Force That Connects Us All.* Oakland, CA: New Harbinger Publications, 2021.

Waite, Amory H. Jr. "Report of the Signal Corps Observer: Operation 'Highjump.'" December 1, 1946–April 1, 1947. http://www.bahaistudies.net/asma/Operation_Highjump.pdf.

Wall, Mike. "NASA Gets $25.4 Billion in White House's 2025 Budget Request." Space.com, March 11, 2024. https://www.space.com/nasa-white-house-2025-budget-request.

Wallach, Michael. *The Viral Delusion* (documentary series). Paradigm Shift, 2022. https://paradigmshift.uscreen.io/.

Wallis, Paul. *Echoes of Eden: What Secrets of Human Potential Were Buried with Our Ancestors' Memories of ET Contact?* Paul Wallis Books, 2022.

Wallis, Paul. *The Eden Conspiracy: Ancient Memories of E.T. Contact and the Bible before God.* Paul Wallis Books, 2023.

Wallis, Paul. *Escaping from Eden: Does Genesis Teach That the Human Race Was Created by God or Engineered by ETs?* Hampshire, UK: John Hunt Publishing, 2020.

Wallis, Paul. *The Scars of Eden: Has Humanity Confused the Idea of God with Memories of ET Contact?* Hampshire, UK: John Hunt Publishing, 2021.

Walsh, Randy. *The Apollo Moon Missions: Hiding a Hoax in Plain Sight Part I.* Randy Walsh, 2018.

Walsh, Randy. *The Apollo Moon Missions: Hiding a Hoax in Plain Sight Part II.* Randy Walsh, 2021.

Ward, Bob. *Dr. Space: The Life of Wernher von Braun.* Annapolis: Naval Institute Press, 2005.

Warren, William F. *Paradise Found: The Cradle of the Human Race at the North Pole.* Amsterdam: Fredonia Books, 2002.

Whiteley, Iya, and Olga Bogatyreva. *Toolkit for a Space Psychologist.* 2018.

Williams, B. J. "Revisiting the Ganzfeld ESP Debate: A Basic Review and Assessment." *Journal of Scientific Exploration 25*, no. 4 (2011): 639–61.

Windbridge Research Center. "Completed Studies." Windbridge.org, n.d. https://www.windbridge.org/research/completed-studies/.

Weill, Kelly. *Off the Edge: Flat Earthers, Conspiracy Culture, and Why People Will Believe Anything*. Chapel Hill: Algonquin Books, 2022.

Weinberg, Steven. *To Explain the World: The Discovery of Modern Science*. New York: HarperCollins, 2015. https://archive.org/details/toexplainworlddi0000wein/page/252/mode/2up?q=proposition+43.

Weiss, David. *Flat Earth Sun, Moon & Zodiac Clock App*. Flat Earth 101 website, n.d. https://www.flatearth101.com/fe-david-weiss.

Weiss, Michael. "Mercury's Orbital Precession, General Relativity, and the Solar Bulge." Updated by Steve Carlip, 1996. *University of California, Riverside* website, n.d. https://math.ucr.edu/home/baez/physics/Relativity/GR/mercury_orbit.html.

Whitrow, G. J. *The Structure and Evolution of the Universe*. New York: Harper & Brothers, 1949 (Reprint, 1959).

Witsit Gets It. "Flat Earth Debate Review: Whitsitt vs. Harrison." YouTube video, April 23, 2024. https://www.youtube.com/live/QDCzNoX1Uzg?si=oSZyfJUEqnRRRJT3.

Witsit Gets It. "In The Field – Robert Bennett Ph.D. – Part 1." YouTube video, November 1, 2023. https://www.youtube.com/live/5j0wCrkzfwk.

Witsit Gets It. "NASA Fraud." YouTube video, February 24, 2024. https://www.youtube.com/live/CJnwl5aiBSk.

Witsit Gets It. "'Professor Dave' vs. Whitsitt – Flat Earth Debate Review." YouTube video, June 17, 2024. https://www.youtube.com/watch?v=qc1oG6yzJYg&t=2995s.

Witsit Gets It. "Relatively Speaking – Episode 8 – Hubble Crisis." YouTube video, June 19, 2023. https://www.youtube.com/watch?v=wSe_GBITdwk.

Witsit Gets It. "Relatively Speaking – Episode 12 – The Death of Modern Cosmology." YouTube video, November 8, 2023. https://www.youtube.com/live/8CEEv4ezJBM.

Witsit Gets It. "NASA Fraud – Episode 3." YouTube video, May 13, 2024. https://www.youtube.com/live/_gpM5lmRVW8.

Witsit Gets It. "Schooling Globers – Episode 7 – Azimuthal Grid of Vision." YouTube video, March 31, 2023. https://www.youtube.com/watch?v=HV89DJhCsCE&t=5672s

Witsit Gets It. "Schooling Globers – Episode 13 – Magnetic Declination." YouTube video, May 31, 2023. https://www.youtube.com/live/2BLIkqri2hg.

Witsit Gets It. "Schooling Globers – Episode 16 – Sonar." YouTube video, June 21, 2023. https://www.youtube.com/live/I-4Wp3e8kIg.

Witsit Gets It. "Schooling Globers – Episode 23." YouTube video, December 13, 2023. https://www.youtube.com/live/7QYaxbsVgd0.

Witsit Gets It. "Schooling Globers – Episode 25." YouTube video, January 10, 2024. https://www.youtube.com/live/3yYdQbKCSC8.

Witsit Gets It. "Schooling Globers – Episode 26." YouTube, January 24, 2024. https://www.youtube.com/live/FBKrYJ0YvUY.

Witsit Gets It. "Schooling Globers – Episode 33 (Geocentrism)." YouTube video, May 1, 2024. https://www.youtube.com/live/oY1svV3dMVg.

Witsit Gets It. "Schooling Globers – Episode 34 (The Core of the Globe Belief)." YouTube video, May 8, 2024. https://www.youtube.com/watch?v=lTIFsHH5i6U&t=2708s.

Witsit Gets It. "Schooling Globers – Episode 35 (Logic)." YouTube video, May 22, 2024. https://www.youtube.com/live/9cRWrHuPYp8.

Witsit Gets It. X post, June 4, 2024. https://x.com/Witsitgetsit/status/1798165881429668322?s=46&t=ijTvU_628uOjAUiS9U8Hxw.

Wolf, Jessica (UCLA). "The truth about Galileo and his conflict with the Catholic Church." UCLA website, December 22, 2016. https://newsroom.ucla.edu/releases/the-truth-about-galileo-and-his-conflict-with-the-catholic-church.

Yale University. "What Is Dark Matter?" COSINE-100 Dark Matter Experiment website, n.d. https://cosine.yale.edu/about-us/what-dark-matter.

York, Herbert. *Arms and the Physicist*. Woodbury, NY: The American Institute of Physics, 1995.

Young, Steven A. *A Fool's Wisdom: Science Conspiracies & The Secret Art of Alchemy*. 2024.

Zeck, Alec. "DEBUNKING THE NONSENSE." The Way Forward Unite Live website, July 21, 2022. https://unite.live/the-way-forward/the-way-forward?recording_id=1110.

INDEX

G

H

I

J

O

P

Y

Z

ABOUT THE AUTHOR

Mark Gober is the author of *An End to Upside Down Thinking* (2018), which won the Independent Publisher (IPPY) award for best science book of the year. He is also the author of *An End to Upside Down Living* (2020), *An End to Upside Down Liberty* (2021), *An End to Upside Down Contact* (2022), *An End to the Upside Down Reset* (2023), and *An End to Upside Down Medicine* (2023), and was the host of the podcast *Where Is My Mind?* (2019). Previously, Gober was a partner at Sherpa Technology Group in Silicon Valley and worked as an investment banking analyst with UBS in New York. He has been named one of *IAM*'s *Strategy 300: The World's Leading Intellectual Property Strategists.*

Gober graduated magna cum laude from Princeton University, where he wrote an award-winning thesis on Daniel Kahneman's Nobel Prize–winning "Prospect Theory" and was elected a captain of Princeton's Division I tennis team.

Made in the USA
Coppell, TX
27 February 2026